690.24 WOO
Woodson, R. Dodge   NOV   2003
REMODELER'S INSTANT ANSWERS
0071398295 $50.00 c2003

R
INS

**DRIFTWOOD LIBRARY
OF LINCOLN CITY**
801 S.W. Highway 101
Lincoln City, Oregon 97367

REMODELER'S INSTANT ANSWERS

R. Dodge Woodson

McGRAW-HILL

New York Chicago San Francisco Lisbon London
Madrid Mexico City Milan New Delhi San Juan
Seoul Singapore Sydney Toronto

The McGraw·Hill Companies

Copyright © 2003 by The McGraw-Hill Companies, Inc. All rights reserved. Printed in the United States of America. Except as permitted under the United States Copyright Act of 1976, no part of this publication may be reproduced or distributed in any form or by any means, or stored in a data base or retrieval system, without the prior written permission of the publisher.

1 2 3 4 5 6 7 8 9 0 DOC/DOC 0 9 8 7 6 5 4 3

ISBN 0-07-139829-5

The sponsoring editor for this book was Larry S. Hager. It was set in Stone Sans by Lone Wolf Enterprises, Ltd.

Printed and bound by RR Donnelley.

This book is printed on recycled, acid-free paper containing a minimum of 50% recycled, de-inked fiber.

McGraw-Hill books are available at special quantity discounts to use as premiums and sales promotions, or for use in corporate training programs. For more information, please write to the Director of Special Sales, McGraw-Hill Professional, Two Penn Plaza, New York, NY 10121-2298. Or contact your local bookstore.

Information contained in this work has been obtained by The McGraw-Hill Companies, Inc. ("McGraw-Hill") from sources believed to be reliable. However, neither McGraw-Hill nor its authors guarantee the accuracy or completeness of any information published herein, and neither McGraw-Hill nor its authors shall be responsible for any errors, omissions, or damages arising out of use of this information. This work is published with the understanding that McGraw-Hill and its authors are supplying information but are not attempting to render engineering or other professional services. If such services are required, the assistance of an appropriate professional should be sought.

**DRIFTWOOD LIBRARY
OF LINCOLN CITY
801 S.W. Highway 101
Lincoln City, Oregon 97367**

*This book is dedicated to Afton and Adam,
the best children a father could ever hope for.
My dedication extends to my father,
who continues to be a driving force in my life,
Victoria who is always supportive, and Nate and Jon,
who have helped give me perspective in the last several years.*

CONTENTS

Introduction   ix

Chapter 1: Basement Conversions 1

Chapter 2: Basement Bathrooms 15

Chapter 3: Flooring and Wallcoverings 33

Chapter 4: Kitchen Remodeling 59

Chapter 5: Bathroom Remodeling 111

Chapter 6: Room Additions 147

Chapter 7: Sunrooms 187

Chapter 8: Attic Conversions 215

Chapter 9: Dormer Installations 243

Chapter 10: Enclosed Porch Conversions 261

Chapter 11: Garages 271

Chapter 12: Decks, Gazebos, and Screened Porches 291

Chapter 13: Cosmetic Remodeling 313

Chapter 14: Dust and Debris Control 321

Chapter 15: Rip-Outs of Existing Conditions 327

Chapter 16: Safety Issues 337

Index 353

INTRODUCTION

Woodson has done it again! He has created yet another book that is an indispensable tool for anyone working within the remodeling industry. If you are looking for a unique resource that provides easy access to nearly instant facts from a seasoned, hands-on expert, this is it. The graphic design of this book makes finding fast answers to your questions simple. Whether in the office or in the field, you can't beat this detailed guide to countless remodeling situations.

Hundreds upon hundreds of tables and figures give you visual reference to a vast array of remodeling scenarios. There are conversion tables, quick-reference tables, graphics, tip boxes, and hard-hitting facts to make your work go faster, smoother, and more profitably.

Woodson has owned his own business since 1979. In addition to his expertise as a remodeler, he has built as many as 60 homes a year and is licensed as a master plumber and master gasfitter. He has worked in the construction industry for over 25 years.

Woodson has written many best-selling books for McGraw-Hill and has served as adjunct faculty for Central Maine Technical College as an instructor of vocational trades. His reputation in the trade is undisputed.

What will you gain from owning this book? More than you can imagine. Take a few minutes to review the table of contents. Thumb through the pages. Look at how easy it is to find the data you need. Where have you seen so much information offered in such a fast, accessible, easy-to-understand format? There is no other remodeling reference like this one. No remodeling contractor or remodeler should be without this valuable tool.

Mr. Woodson would like to take this opportunity to thank the United States Government for providing numerous pieces of artwork for this book.

Some images © 2002--www.arttoday.com

Chapter 1

BASEMENT CONVERSIONS

Basement conversions are a popular type of remodeling. Homeowners often turn to this type of remodeling to create space at a minimum cost. It is true that the cost per square foot for a basement conversion is one of the most economical ways to create livable space. However, it is rarely a good investment. Trying to recover the cost of a basement conversion on a real estate appraisal can be difficult. At best, a property owner is likely to break even on the expense. When compared to kitchen or bathroom remodeling, where the investment can result in a paper profit of appraised value, basement conversions are not wise. Even so, many people turn to these conversions when they need more living space.

If a person is seeking additional space at an affordable cost and is not intent on seeing a strong return on the investment, a basement conversion can be just the ticket. Remodeling contractors face specialized needs when working with basements. Ceiling height can be a problem. Support posts and beams can present trouble for a remodeler. Dampness is another common problem with basement conversions. Once the basic basement problems are dealt with, the remodeling work itself can be fairly easy.

TYPES OF BASEMENTS

There are three basic types of basements. Each type can require different remodeling tactics. A walk-out basement offers the most options for a viable basement conversion. Daylight basements rank second behind walk-out basements. A buried basement comes in last, but it still has potential for a remodeling conversion. Walk-out basements and daylight basements offer the advantage of natural light. This can be a major factor in making remodeling decisions. The light can affect the types of wall coverings used and the feel of an expansive habitat. Buried basements tend to be more damp than the other types of basements that have better ventilation. Remodelers work with all three types of basements and must adjust their procedures to meet the needs of the space that is being converted into living space.

Walk-Out Basements

Walk-out basements are equipped with legal egress doors. The door in a walk-out basement is often a sliding-glass door, but it can be a terrace door or a standard entry door. Part of the basement is likely to be buried, but at least one portion of the basement will be above ground, so that a person can step out of the basement onto grade level.

Did You Know: that fire codes come into play when basement conversions including sleeping areas? Check your local codes to confirm what means of egress are required for bedrooms in basements. You are likely to find that buried basements will not be approved to house bedrooms.

Don't Do This! If you are converting a buried basement into living space, avoid using dark colors. Stay away from dark paneling or paint. Natural light is nearly nonexistent in buried basements, so use light colors to brighten the area and to give a greater feeling of expansive space.

Daylight Basements

Daylight basements have windows in them, but no exterior doors. It is common for the basement walls to rise several feet below grade and then to turn into wood framing that contains windows of normal sizes. The windows can be double-hung, casement, or sliders. These windows provide both light and ventilation. This is a valuable asset when considering a basement conversion.

Buried Basements

Buried basements are the least appealing for conversion purposes. These basements are buried in the earth and have little more than tiny windows at the highest portion of the interior walls. Converting a buried basement results in a dark living space if creative lighting and wall coverings are not used.

PLANNING

Planning a basement conversion is an important step in the remodeling process. Homeowners will often know what they want to do. It is up to the remodeling contractor to determine what can be done in a reasonable manner. Property owners frequently have good ideas, but it is not uncommon for them to have unrealistic expectations. Using a checklist to evaluate a basement is a good place to start the planning phase. It is best to create your own checklist from past experiences and to update it as you learn more from additional jobs. You might want several checklists to be sure that you have as many bases as possible covered. Going over a checklist with your customers will give you an air or professionalism and reduce the likelihood of missing a key factor in your bidding and planning process. See the following checklist as a short sample of what your checklist might include.

Basement Evaluation Checklist

- ✔ What are the means of ingress and egress?
- ✔ What is the available ceiling height?
- ✔ Are there plumbing pipes of heat ducts hanging below the ceiling joists?
- ✔ Where is the mechanical equipment located?
- ✔ Are there signs of water leaks or dampness?
- ✔ What condition is the existing floor in?
- ✔ Are the existing walls made of concrete or block?
- ✔ Do support posts limit floorplans?
- ✔ Is there plumbing roughed into the existing floor?
- ✔ Is there adequate expansion room in the electrical panel?
- ✔ How will the basement conversion be heated?
- ✔ What is required for stairs that will meet code requirements for living space?
- ✔ How will remodeling debris be removed from the basement?
- ✔ How will building materials be brought into the basement?

The checklist above is representative of the types of questions you may want to ask yourself when evaluating a basement conversion. Of course, the list is not all inclusive, but it does represent the types of questions to consider putting on your own checklist. Working with a checklist can reduce your level of frustration and the risk of losing money by overlooking key elements in the bidding process.

WATERPROOFING

Waterproofing might be necessary before converting a basement into living space. This can be as simple as applying a waterproofing material to a basement wall or as complicated as installing drain tile around the perimeter of the basement, installing a sump crock and a sump pump.

Trade Tip: When checking a basement for a possible conversion project, sketch existing conditions in a notepad. Pay attention to support posts, beams, ducts, plumbing, heating systems, laundry facilities, and other elements that might restrict the use of a particular portion of the basement. Having such a sketch on file can be of great help when drafting a working floor plan for the remodeling project.

Basements with extreme water infiltration should be retrofitted with drain tile. In extreme cases, this should be done outside of the foundation wall and inside the foundation. Most basements can be dried up with an interior installation. The process involves breaking up the perimeter edges of the concrete floor to allow the installation of slotted drain pipe. Plastic, slotted drain pipe should be bedded in and covered with crushed stone. The pipe is not expensive, but the demolition and repair of the floor can be, so don't overlook this step if you see evidence of water marks on the foundation walls during your estimating inspection. The pipe is run into a sump crock that is installed below floor level. A sump pump empties the sump when the water reaches a certain level. Water coming into the foundation is captured by the slotted pipe. If the pipe is installed properly and graded to deliver the water to the sump, all of the foundation water will be diverted into the sump and pumped out of the

Did You Know: that some basement have natural streams running under them? They can. During my years as a remodeler, I have run into underground running water more than once when installing basement bathrooms. In the early going, I wasn't aware of this risk. As I gained experience, I put a disclaimer in my contracts that allowed me to charge additional fees if underground running water was encountered when a basement floor was opened for an installation.

home. This type of action is rarely needed but it can be, so don't get sloppy during your inspection of the basement.

There are a number of waterproofing mixtures on the market that can be applied to the surface of foundation walls. In some cases, the mixtures are applied to specific areas of a wall, such as an obvious crack and leak. Other cases require the covering of full foundation walls to prevent dampness. You will have to assess each basement wall individually and match the proper material to your personal need.

STAIRS

The stairs in basements often don't meet the code requirements for stairs used in general living space. Check your local code requirements and make sure that any existing stairs are suitable. You may find that you will have to increase the head space of the staircase. Another possibility is that the stair rise will be too great to meet code requirements for living space. If you have to extend the stairs to obtain a suitable rise, you will lose floor space in the basement. This is information you need to know before you begin the remodeling process.

WALL PREPARATION

Wall preparation for foundation walls can take on a few forms. It's not feasible to install finished wallcoverings directly to a foundation wall. Therefore, you must decide whether to use furring

 Trade Tip: Most building codes require an average ceiling height that is not less than seven feet six inches in height. Confirm this with your local code and be aware of height requirements before you begin remodeling a basement. The last thing you need is to complete a beautiful basement conversion only to find that you cannot get an occupancy permit for it due to inadequate head space.

strips or full wall studs. Furring strips are less expensive, but they don't offer as much flexibility as full-sized wall studs do. It is likely that the walls will be insulated. With furring strips, only minimal insulation is possible. Full-sized wall studs allow for complete insulation. Electrical wires and plumbing are also considerations. It is much better to have stud walls to work with when wiring and plumbing is desired in the exterior walls. Another advantage to building a stud wall is the ability to plumb it up. Furring strips used on a foundation wall that is out of plumb makes the entire job of finishing the living space difficult. All in all, I suggest you work with wall studs instead of furring strips.

FLOORING

The flooring in a basement can be as simple as painted concrete to something as nice as a beautiful hardwood floor. Carpet is the most common choice of flooring in basement conversions. When carpet, vinyl, or wood flooring is desired, the concrete floor must be prepared to accept the finished floor covering. There are different ways to prep a floor.

One method of preparing a floor is to install a vapor barrier of plastic over the concrete that will be covered by 4 x 8 sheets of plywood. Another method is to install treated lumber over a vapor barrier to create an airgap between the plastic and the floor sheathing. In this type of situation, plywood is nailed to the treated lumber. This method is best if the existing floor is not level. The treated lumber allows an installer the opportunity to level the floor prior to installing final floor sheathing. Once the sheathing, also known as subflooring, is in place, the remainder of the flooring installation can proceed as it would with any other flooring installation. See Chapter 3 for more comprehensive flooring details. What follows are several flooring options:

- Carpet
- Vinyl
- Quarry tile
- Ceramic tile
- Wood tile
- Hardwood strip flooring
- Softwood board flooring

Don't Do This! Do not install particle board or wafer board directly on a concrete basement floor. These materials are likely to wick up moisture and warp.

SUPPORT COLUMNS

Support columns are common in basements (Figure 1.1). These columns can cause considerable problems when it comes to creating a desirable floor plan. One approach to take with columns is to box them in. This makes the posts more attractive, but it does not open up the floor space. There is an alternative in most cases. While you may not be able to eliminate all columns, you can certainly reduce the number of them when you install a strong beam. This beam will probably be either a steel beam or a wood girder with a steel flitch plate in it to support the floor above the basement.

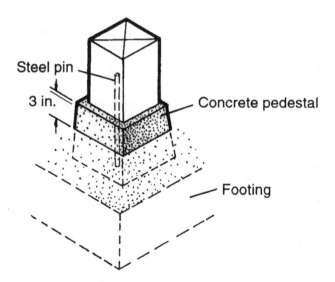

FIGURE 1.1 A typical basement support column on a concrete pedestal.

Trade Tip: Eliminating support columns does wonders in opening up the usable space of a basement. The installation of a suitable support beam will often eliminate the need for most, if not all, support columns.

Jack posts are sometimes installed after a home is built to correct or prevent the sagging of floor joists. These posts usually sit on the surface of the concrete floor (Figure 1.2). If you are going to install a beam to eliminate primary support columns, you should also be able to remove jack posts. You will, of course, have to keep the joists supported with temporary jacks until the new beam is in place.

INTERIOR FRAMING

Interior framing for a basement conversion is no different than the interior framing used in routine building and remodeling practices. The same principles are applied to non-bearing, interior framing in a basement as you would use in any other residential application (Figure 1.3).

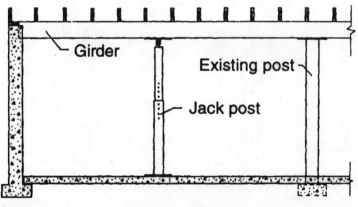

FIGURE 1.2 Jack post being used as additional joist support.

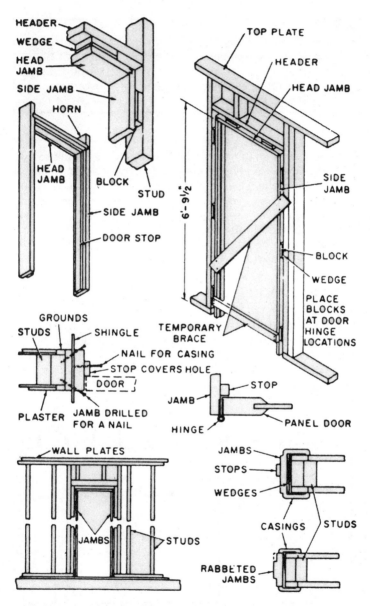

FIGURE 1.3 Typical framing for a door jamb and trim.

CEILINGS

Finished ceilings in basements can be as simple as drywall applied to ceiling joists. This is often the most common form of ceiling coverings. There are times, however, when a suspended ceiling is preferable. This is true most often when there is substantial plumbing, wiring, or heating systems installed below the existing joists. Either type of ceiling is fine.

WALL COVERINGS

Wall coverings in basements range from drywall to sheet paneling to wooden planks. Drywall is usually the most cost effective wall covering to use. The wall covering options are the same as what would be used in other parts of a home. See Chapter 3 for more details on these options.

Sequence of Events During a Typical Basement Conversion

- ✔ Waterproofing
- ✔ Framing
- ✔ Installation of rough plumbing, heating, and electrical
- ✔ Insulation
- ✔ Wall coverings
- ✔ Ceilings
- ✔ Paint
- ✔ Finish flooring
- ✔ Cabinets
- ✔ Trim
- ✔ Final plumbing, heating, and electrical
- ✔ Test systems
- ✔ Final inspection

Trade Tip: Basements tend to be dark, so choose wall coverings that are light in color.

QUESTIONS AND ANSWERS

Q: Should I insulate the ceiling joists in a basement before installing a celing?

A: *Not normally. It is sometimes done for noise reduction, but there is no advantage to insulating the ceiling joists beyond noise control.*

Q: If I have a basement that could accept a full-size entry door but that is not built for one, can I cut out a section of the foundation wall to install a door?

A: *You should be able to. A concrete saw will take care of the masonry wall. A suitable header will be needed over the door opening to support the floors above. Based on past experience, I recommend hiring a concrete cutting company to create the opening. Cutting thick masonry walls with a rented saw can result in a sloppy cut.*

Q: I have a daylight basement and am not sure how to make the transition from the masonry wall to the framed wall look good. Do you have any ideas?

A: *This is a common problem. One of my favorite ways is to install wainscoting on the masonry wall and blend it into the sill where the framing begins with the equivalent of a chair rail. A nice wood-boxed sill and face trim makes an attractive transition.*

Q: Should I recommend the installation of a sump-pump system for all basements before finishing them?

A: *That is probably overkill. If a basement has been dry for years, it probably will not develop leaks of a significant nature. If I were building a home with a basement, I would rough it in with a sump-pump system, but doing so for a remodeling job adds a good deal of expense to the project.*

Q: Why can't I install vinyl flooring directly to the concrete floor of a basement?

A: *You can, but you may be buying yourself a lot of trouble. Without any air space between the vinyl and concrete, the vinyl may not lay flat during use. This would not be a problem with ceramic or quarry tile.*

Q: Should I use treated lumber for my base plates on wall partitions when the wood is in direct contact with concrete?

A: *It's a very good idea to do so.*

Q: How difficult is it to put a bathroom in a basement that does not have any plumbing roughed-in for one?

A: *It is a substantial job under any conditions. If the building drain or sewer is under the floor, the job is very feasible. If the building drain or sewer leaves the building above the floor level, a sewer-ejector system will be needed. This is not particularly difficult to install, but it is costly.*

Chapter 2

BASEMENT BATHROOMS

Basement bathrooms are popular in many areas, and there are a number of contractors who rely on the installation of basement bathrooms for much of their annual income. Some basements are roughed-in for the future installation of bathroom fixtures. Other basements have no pre-planned provisions for plumbing at the floor level.

A basement that has not been roughed-in can present numerous challenges for the remodeling contractor who is asked to install a bathroom.

Each job comes with its own set of circumstances. Some jobs are fast and easy. Others are complex and can present unusual problems. My crew once broke up a basement floor to install drainage piping for a basement bathroom. What they found was a shock to all of us. There was an active, running stream under the floor of the home. There was so much water, moving so quickly, that the use of plastic pipe was out of the question. We had to install cast-iron drains to combat the water. I've installed a lot of basement baths and this is the only time I have ever encountered so much water. But water isn't the only potential obstacle.

It is common to find basements that do not have a building drain installed below the floor level. Building drains often leave homes through the wall of a basement. When this is the case, a sump and sewage ejector pump is needed. This adds considerably to the cost of a job, so the need for a pump must not be overlooked.

If you are going to bid a job for the installation of a basement bath, there are several key features to consider. What should you look for? Check out the following list:

Key Factors to Consider for Basement Baths

✔ Are drains roughed in for plumbing?
✔ Are vents roughed in for plumbing?
✔ What is the available access for installing a bathing unit?
✔ Will a sewage sump and pump be required?
✔ How much concrete will need to be broken and repaired?
✔ How far away are adequate water supplies?
✔ Is the existing sewer large enough to support an additional bathroom?
✔ What is the available routing for vents that may be needed?
✔ How far is the proposed bath location from the existing building drain?
✔ Will the existing electrical service handle the new load of an additional bathroom?
✔ What is the available routing of electrical wiring?
✔ How difficult will it be to provide exhaust-fan ventilation?

The checklist above is not conclusive of all questions that you will need to ask yourself, but it is representative of the types of concerns you should consider. Do your homework before you make your bid. Getting into a job only to discover that you need a pumping station that you didn't plan on is an expensive error.

ROUGHED IN

When you're fortunate enough to have a basement where plumbing is already roughed in, your job will be much easier. There are different levels of rough-in. You must look carefully to determine how much of the work has been done for you and your plumbing contractor. Start by looking for drains that protrude from the floor. Assuming that you are dealing with a full bath, you will need a 3 inch drain for a water closet, a 2 inch drain for a lavatory, and a tub box that contains a 2 inch drain for either a shower or bathtub. While bathtubs and lavatories are piped above ground with 1½ inch pipe, any pipe installed below concrete is required to have a minimum diameter of 2 inches. Check on this, because you could have trouble getting your job approved by a plumbing inspector if the rough-in is not sized properly.

In addition to drains, a basement bathroom requires venting. A minimum vent size when a toilet is installed is 2 inches. This is assuming that there is a 3 inch vent somewhere else in the building. The vent may tie into other vents at higher levels or may extend through the roof to open air. Vents that tie into other vents are required to make their connections at levels well above the flood-level rim of existing fixtures. This is normally about 42 inches above the floor level where the fixture is located. If you have to run a vent from a basement to an attic, the routing can be tricky.

Vents are sometimes roughed in for future basement baths. When this is the case, you should see a 2 inch pipe extending into a joist bay in the ceiling of the basement. If there is a capped pipe waiting for a vent connection from the basement bath, you've got it made.

Water pipes are not normally run to fixture locations in roughed-in baths. However, there should be water distribution pipes of adequate size nearby when a future-use bath location has been roughed in. It is standard procedure to put no more than two fixtures on a half-inch water supply. In a full bath, you

 Did You Know: that the minimum pipe diameter for drainage pipe installed below concrete is 2 inches? This is a code requirement, so don't work with or install pipe of a smaller diameter when it will be covered with a concrete floor.

Trade Tip: Do you need a plumbing permit to install fixtures in a basement bath that is already roughed in? Yes, you do. Under normal conditions, the rough-in was expected during the general construction phase of the initial building activity. Even so, you are required to obtain a permit and an inspection for the installation of fixtures.

will need two hot water supplies and three cold water supplies. This means that you should have a ¾ inch pipe available for cold water and a ½ inch pipe available for hot water.

NO ROUGH-IN

What do you do when there is no rough-in available for a basement bath? You earn your money. The requirements are the same as those discussed above, but getting the piping to where you need it is much more difficult. A jackhammer will be needed to break up the existing concrete floor to allow the installation of piping. My crews have used electric jackhammers very successfully for this purpose.

Before you begin tearing out a floor, you must decide where the piping will need to begin and end (Figure 2.1). Are you tying into an existing building drain below the floor? Will you need to install a sump and a sewer ejector pump? This is one of the first considerations. In order to tie into an existing drain below grade, you will have to locate the pipe. If the basement is unfinished, this is fairly easy. You simply look for where a suitable drain enters the floor and go to it. Finished basements are not so easy to pinpoint drains in.

Did You Know: that bathrooms must be equipped with electrical circuits that are ground fault protected? They are. You can do this with a ground fault outlet or with the use of a ground-fault breaker in the circuit breaker panel.

Don't Do This! Don't start breaking up a basement floor for piping from the bathroom location. Why? The building drain may not be deep enough to allow your plumbing contractor to install the new drain with proper grade on it. There is also a risk that the building drain will be too small to accept a new bathroom. Break up the floor at the building drain location first. Then you will know what you are dealing with and if you can't use the pipe, you will have considerably less floor to repair.

To find the building drain in a finished basement, go outdoors. Yes, I said to go outdoors. Look around the foundation of the building for a cleanout plug. It should be within five feet of the foundation. Once you find the cleanout plug, you will know where the building drain leaves the building. Measure from the corner of the building to the cleanout and then go to the basement and measure from the same corner of the foundation to find the location of the building drain. Simple, huh? But you are not done, yet.

Okay, you found the building drain. Now you have to find out if it is deep enough to do you any good. It is common practice for plumbing drains to have a grade of ¼ inch of fall for every liner foot the pipe travels. In other words, a 4 foot length of pipe will be one inch lower at its downstream end than it is at its upstream end. You must confirm that the building drain is low enough in the floor to allow a suitable connection. Let's say that the new bath location is 20 feet from the connection point. A plumber will need to start the drainpipe for the bathroom at a point 5 inches above the location of the building drain. If the building drain is only 3 inches below the floor, you've got a problem. Beyond this, room is needed for an elbow to turn up for the drain of a water closet. This requires a few extra inches. In reality, the building drain would probably have to be about 8 to 10 inches lower than the drain for the bathroom where it begins.

Once you have a suitable drain to connect to, you can break up the existing floor and install the below-grade piping. Remember to have this piping inspected by the local plumbing inspector before you conceal it. Your plumbing contractor should take care of this, but make sure that you have an approved inspection slip before your crew replaces the floor.

Trade Tip: If copper pipe is run below concrete, it must be sleeved where it comes up out of the concrete. Copper and concrete don't get along well together. If you are running water pipes below grade using copper tubing, make sure that you install foam insulation or some other suitable sleeve when you turn the tubing up out of the floor to supply fixtures with water.

Venting

Venting a basement bath that is not roughed in can be difficult. There are a few options. A 2 inch vent is required. One option is to find an existing bathroom on the floor above the basement and open the wall to allow a vent to be tied into the vent for the existing bathroom. This creates the costly need of repairing the wall.

I have been successful in working vents from basements to attics without opening walls, but you can't count on this option. If fire blocking is in the existing partition walls, you can't do this. There is also a chance that existing electrical wiring will block the

FIGURE 2.1 A typical rough-in of below-grade piping for a basement bathroom.

installation of a pipe in a concealed wall. The way to do this, when it can be done, is to locate walls with precise measurements. From the basement, drill through the floor above and through the base plate of the partition. Make sure the hole is large enough to accept the diameter of a pipe coupling. The pipe will have to be put through the wall in sections, so couplings will be needed. Go to the attic and drill a hole that will line up with the hole in the base plate. This is not an easy task, but it is possible. Once you have both holes, drop a plumb bob from the top hole to make sure that the holes line up. The next step is to snake the pipe through the hole. If you don't encounter wiring, ductwork, fire blocking, or other obstacles, you will be successful.

Another option, if the property owner doesn't mind the appearance, is to run the vent up the outside of the building. You can exit the basement wall and run the vent vertically to a point above the roofline. When this is done, the exterior piping must be protected from freezing by running an oversized pipe or by insulating it and boxing it in. Normally, this is a last resort.

Water Piping

Water piping is usually the easiest part of plumbing a basement bath. When working in an unfinished basement, getting water pipe to the proper location is rarely a problem. Finished basements are another story. Your options in a finished basement are limited. One way to conceal piping is to install it below the floor.

 Don't Do This! Don't connect the drain of a basement bath to a 3 inch drain that already has two toilets connected to it. A 3 inch pipe is not allowed to carry the waste of more than two toilets. This can be a serious problem, especially if a building has two existing toilets and only a 3 inch sewer. When this is the case, the building is legal as it is, but the addition of another toilet will be illegal. To make the installation, a new building sewer will have to be run to the main sewer where a larger drain is available for connection. Falling into a trap like this can erase any hope you had of making a profit on a job.

Another way is to run the piping near a beam or wall and box it in. Short of these options, the finished ceiling must be opened for the pipe installation and then repaired later. Remember that all concealed piping must be inspected by the local plumbing inspector before it is concealed.

FIXTURE PLACEMENT

Fixture placement for a basement bath must meet the same requirements as those used for any other typical bathroom (Figure 2.2). This is a matter for your plumbing contractor, but it is wise to check the pipe placements before you seal the floor. Again, make sure that you have an approved inspection slip prior to repairing the concrete floor.

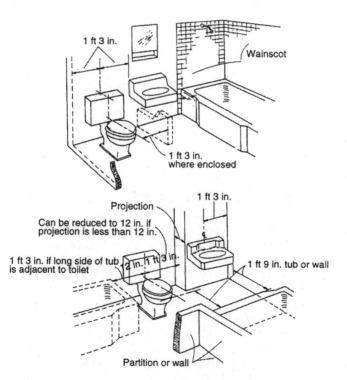

FIGURE 2.2 Typical spacing for plumbing fixtures.

SEWAGE EJECTORS

Sewage ejectors are needed for basement baths when drains cannot be installed to drain by natural gravity. When installed for a basement bath, these systems are not complex. The basic procedure is to break open the floor, dig a hole, and install a sump basin (Figure 2.3).

The pumps used to empty a sewage sump are easily installed. Basically, the pump is set into the base of the sump, piped up, and plugged in. They operate with a float that turns the pump on when the level of contents in the sump reaches a certain point. Once the contents are emptied, the float triggers the pump to turn itself off (Figure 2.5, 2.6, 2.7).

Your plumbing contractor should take care of sizing the sewage-ejector system for a basement bath (Table 2.1). There are code requirements on the sizing. As a remodeler, you are not expected to be a master plumber, but don't fail to make sure that the pump used on your job is within code compliance.

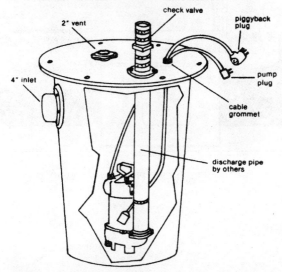

FIGURE 2.3 Typical sewer ejector setup. *(Courtesy of A.Y. McDonald Mfg. Co.)*

24 Remodeler's Instant Answers

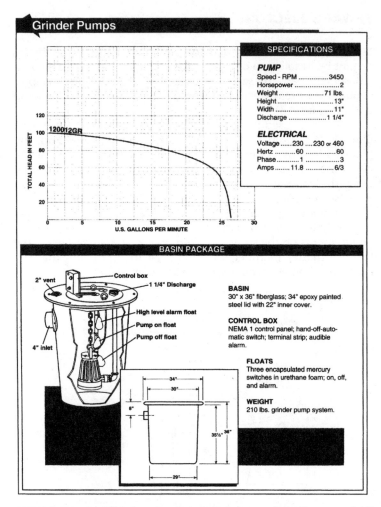

FIGURE 2.4 Specifications for a typical sewage sump. *(Courtesy of A.Y. McDonald Mfg. Co.)*

Grinder Pumps

The **A.Y. McDonald Mfg. Co. Grinder Pump** comes as a completely packaged simplex system. It is designed for residential and small industrial sewage or sump applications. The pump is recommended for homes in isolated or mountainous areas, and for dewatering of dwellings located in inland protected areas where septic tanks are not permitted. System includes pump, tank, cover, check valve, discharge piping, control box and float switches.

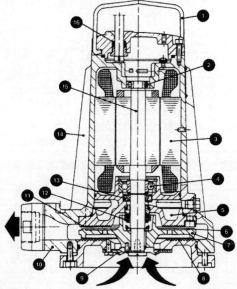

FEATURES

1. Lifting handle
2. Self-lubricating ball bearing-needs no servicing
3. Stator, insulated against heat and humidity to class F (155°C)
4. Oversized single row ball bearing
5. Oil chamber for lubrication and cooling the seal assemblies
6. Back vanes on impeller
7. Dynamically balanced impeller
8. Adjustable spiral bottom plate for handling fibrous material
9. Patented hardened rotor and stator cutter elements (Rockwell C 58-62)
10. Volute with centerline discharge suitable for mounting to guide rail bracket or discharge elbow
11. Spiral back plate
12. Mechancial lower seal enclosed in Buna N boot
13. Upper lip seal angle mounted for long life
14. Motor housing with large cooling fins
15. Rotor shaft assembly dynamically balanced
16. Watertight cable joint with strain relief

FIGURE 2.5 Sewage-ejector setup with an alarm system. *(Courtesy of A.Y. McDonald Mfg. Co.)*

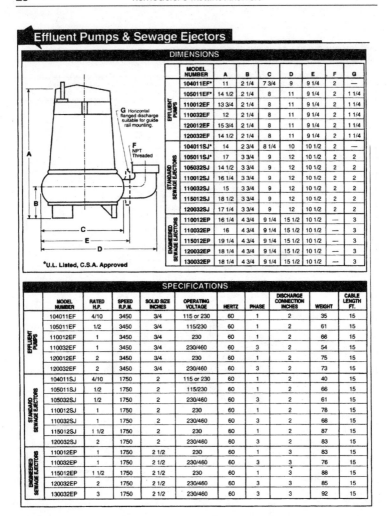

FIGURE 2.6 Specifications for effluent pumps and sewage ejectors. *(Courtesy of A.Y. McDonald Mfg. Co.)*

Basement Bathrooms

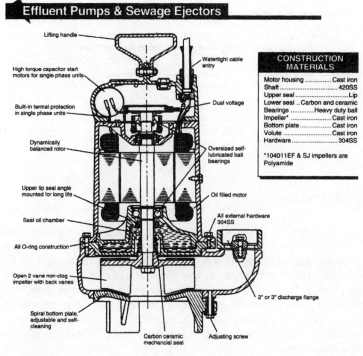

FIGURE 2.7 Cutaway of a submersible effluent pump. *(Courtesy of A.Y. McDonald Mfg. Co.)*

TABLE 2.1 Discharge pipe diameter is compared to the capacity of a pump or ejector. *(Courtesy of International Code Council, Inc. and International Plumbing Code 2000)*

MINIMUM CAPACITY OF SEWAGE PUMP OR SEWAGE EJECTOR

DIAMETER OF THE DISCHARGE PIPE (inches)	CAPACITY OF PUMP OR EJECTOR (gpm)
2	21
$2^1/_2$	30
3	46

For SI: 1 inch = 25.4 mm, 1 gpm = 3.785 L/m.

TABLE 2.2 This table shows sizes for fixture drains and traps, based on fixture units. *(Courtesy of International Code Council, Inc. and International Plumbing Code 2000)*

DRAINAGE FIXTURE UNITS FOR FIXTURE DRAINS OR TRAPS

FIXTURE DRAIN OR TRAP SIZE (inches)	DRAINAGE FIXTURE UNIT VALUE
1 1/4	1
1 1/2	2
2	3
2 1/2	4
3	5
4	6

For SI: 1 inch = 25.4 mm.

PARTITIONS

The installation of partitions for a basement bath usually goes smoothly. Framing the walls and door openings is your standard, cookie-cutter framing. It is a good idea when a bathtub is to be installed to have the tub in the general installation area before you frame the walls. Before you secure the partitions in their final position, check the rough-in measurements of your plumbing pipes for code compliance. It is much easier to adjust the wall locations during the framing stage than it is to move the piping at a later stage.

ELECTRICAL WIRING

The electrical wiring for a basement bath tends to be simple (Figure 2.8). You will need a ground-fault-interrupter (GFI) circuit in the bathroom. If the bathroom doesn't have a window that opens to open air, you will need an exhaust fan. It is likely that some type of wall-mount heat will be installed, so this is another consideration for the electrical rough-in. In general, the wiring is pretty straightforward.

Electrical Considerations

✔ Check to make sure that the existing electrical panel has enough room to add a new circuit for the bathroom you are adding.

Basement Bathrooms

FIGURE 2.8 Typical wiring in a partition wall. Remember to use a GFI circuit in the bathroom.

- ✔ Decide what route the electrical wiring will take.
- ✔ Will you run wiring below the floor?
- ✔ Will you have to box in a chase for the wiring?
- ✔ You must install a GFI circuit in a bathroom.
- ✔ You must install an exhaust fan if the bathroom will not have a window that opens to open air.
- ✔ Will an electric wall heater be installed?
- ✔ Will a heat lamp be installed?
- ✔ What type of light fixtures will be used?

INSPECTIONS

Inspections are required for plumbing and electrical work before the rough-ins are concealed. It is up to the subcontractors to take care of their own inspections, but the general contractor should require copies of approved inspection slips prior to concealing plumbing or wiring.

THE REST OF THE JOB

The rest of the job in finishing a basement bath is essentially the same procedures used for normal building and remodeling work. You can find details on the completion elements in various chapters of this book. For example, the next chapter deals with flooring and wallcoverings. You can refer to that chapter for information on walls and floors.

Plumbing is the most critical and difficult part of adding a basement bathroom. The key to success is to be thorough in your planning process. Don't bid or begin the job until you have a complete plan in mind. Basement bathrooms tend to offer profitable work for remodelers and should not be overlooked.

QUESTIONS AND ANSWERS

Q: I have heard that homes that use sewage ejectors must be equipped with alarm systems. Is an alarm system required for a basement bath when it is the only bathroom group emptying into a sewage sump?

A: *No. The alarm system is not normally needed unless the primary plumbing system empties into a sewage sump.*

Q: Will particleboard vanities warp if they are used in a basement?

A: *It's possible for particleboard vanities to warp, but they usually don't, unless there is significant water damage to them.*

Q: I've got to jackhammer a trench in a basement floor for my plumbing contractor. The problem is that I can't get a cement truck into a position to deliver concrete into the basement to repair the trench. What should I do?

A: *There are two possibilities. You can mix the cement on site with either a portable mixer or a mixing tub. If you have enough workers, you can use either a wheelbarrow or bucket brigade to haul cement into the basement from the cement truck. Neither way is fun, but both will work.*

Q: I can't see a way to get a regular vent out of the basement for the bathroom. Can I use these mechanical vents I have heard about?

A: *You might get a code variance to use a mechanical vent on a sink or lavatory, but it is highly unlikely that any code enforcement office will give you permission to use one for a toilet or a full bathroom group.*

Q: Is ceramic tile a good choice for a floor in a basement bathroom?
A: *Yes.*

Q: Does a basement bath require an electric ventilation fan if there is a foundation window in the bathroom?
A: *Generally, if a bathroom is equipped with a window that opens, a ventilation fan is not required.*

Chapter 3

FLOORING AND WALLCOVERINGS

Flooring and wallcoverings are top choices when it comes to remodeling. Many people like to repaint their homes, add wallpaper, and switch carpeting. Most homeowners are not going to call a remodeling contractor when thinking about replacing the carpeting in a home. More likely, the homeowner will go to a flooring store to purchase the carpeting and installation services. The same can be true when it comes to giving the walls a new look. Yet, property owners frequently call remodeling contractors when the desire is to revamp a room entirely.

There is good money to be made when installing carpeting, tile, and wood flooring. Painting is also profitable. Since both of these tasks become a finished product that people see daily, the work must be done properly. To make the most money on these types of jobs, a contractor has to be aware of the many options that can be offered to customers.

Did You Know: that the use of oil-based paints has dropped dramatically over the last several years? Some of the reasons that this type of paint fell from popularity are that it dries slowly, produce strong odors, and is quite flammable.

Choosing the right type of wallcovering can have a massive impact on the finished look of a remodeling job. For example, installing a dark, wood paneling in a buried basement is a bad idea. Buried basements are dark enough on their own. Installing a dark paneling will only make the darkness problem worse. A buried basement should be finished with light colors.

Would you recommend installing wood flooring in a bathroom? I've seen this done, but I don't think it makes much sense. If the wood is not well protected, it can turn black from exposure to water and dampness that is common in a bathroom.

Is carpeting the best choice for finished flooring in a kitchen? Not usually. Grease spatters, spills of all types, and heavy traffic in a small area will take its toll on carpeting. Vinyl flooring or tile is generally considered to be a better choice. It is information like this that you need to be knowledgeable of.

Property owners are likely to ask you advice when it comes to flooring and wallcoverings. If you don't know how to advise them, you will not gain the great reputation that you would like to enjoy as a remodeling contractor. You might subcontract all of your painting work out to independent contractors. This is not an excuse for not knowing the difference between a latex paint and an alkyd paint.

Before we dive into the depths of this chapter, let's cover a couple of safety issues to keep in mind. As a remodeler, you are likely to be working with older building products. Some of these products may contain health hazards. You could easily run into lead-based paint in existing buildings. Old vinyl flooring may contain asbestos. Vinyl wallpaper from the old days may contain asbestos. Evaluate what you will be working with before you

 Trade Tip: Use alkyd paint when you need a strong surface or superior hiding power.

expose yourself, your crews, and your customers to potential health hazards (Figures 3.1, 3.2).

DRYWALL

Drywall is the most popular type of wallcovering in use today. It is very versatile, relatively inexpensive, and installs quickly. All things considered, drywall is the primary choice of contractors when it comes to covering up wall studs and ceiling joists.

PANELING

Paneling is normally used as an accent wallcovering. Rarely will you find a house that is finished entirely with paneling. Cheap paneling can prove to be problematic over time. It can warp and cause a contractor significant problems in customer satisfaction.

Product	Location	Percent asbestos	Dates of use	Binder	Friable/nonfriable	How fibers can be released
Roofing felts	Flat, built-up roofs	0–15	1910–present	Asphalt	Nonfriable	Replacing, repairing, demolishing
Roof felt shingles	Roofs	1	1971–1974	Asphalt	Friable	Replacing, demolishing
Roofing shingles	Roofs	23–32	?–present	Portland cement	Nonfriable	Replacing, repairing, demolishing
Roofing tiles	Roofs	20–30	1930–present	Portland cement	Nonfriable	Replacing, repairing, demolishing
Siding shingles	Siding	12–14	?–present	Portland cement	Nonfriable	Replacing, repairing, demolishing
Clapboards	Siding	12–15	1944–1945	Portland cement	Nonfriable	Replacing, repairing, demolishing
Sprayed coating	Ceilings, walls, and steelwork	1–95	1935–1978	Portland cement, sodium silicate, organic binders	Friable	Water damage, deterioration, impact
Troweled coating	Ceilings, walls	1–95	1935–1978	Portland cement, sodium silicates	Friable	Water damage, deterioration, impact
Asbestos-cement sheet	Near heat sources such as fire-places, boilers	20–50	1930–present	Portland cement	Nonfriable	Cutting, sanding, scraping
Spackle	Walls, ceilings	3–5	1930–1978	Starch, casein, synthetic resins	Friable	Cutting, sanding, scraping
Joint compound	Walls, ceilings	3–5	1945–1977	Asphalt	Friable	Cutting, sanding, scraping
Textured paints	Walls, ceilings	4–15	?–1978		Friable	Cutting, sanding, scraping
Millboard, rollboard	Walls, commercial buildings	80–85	1925–?	Starch, lime, clay	Friable	Cutting, demolishing
Vinyl wallpaper	Walls	6–8	?		Nonfriable	Removing, sanding, dry-scraping, cutting
Insulation board	Walls	30	?	Silicates	Friable	Removing, sanding, dry-scraping, cutting
Vinyl-asbestos tile	Floors	21	1950–1980?	Poly(vinyl)chloride	Nonfriable	Removing, sanding, dry-scraping, cutting
Asphalt-asbestos tile	Floors	26–33	1920–1980?	Asphalt	Nonfriable	Removing, sanding, dry-scraping, cutting
Resilient sheet flooring	Floors	30	1950–1980?	Dry oils	Nonfriable	Removing, sanding, dry-scraping, cutting
Mastic adhesives	Sheet and tile backing	5–25	1945–1980?	Asphalt	Friable	Removing, sanding, dry-scraping, cutting
Cement pipe and fittings	Water and sewer mains	20–?	1935–present	Portland cement	Nonfriable	Demolishing, cutting, removing
Block insulation	Boilers	6–15	1890–1978	Magnesium carbonate, calcium silicate		Friable damage, cutting, deterioration
Preformed pipe wrap	Pipes	50	1926–1975	Magnesium carbonate, calcium silicate		Friable damage, cutting, deterioration
Corrugated asbestos paper High temperature	Pipes	90	1935–1980?	Sodium silicate, starch	Friable	
Moderate temperature	Pipes	35–70	1910–1980?	Sodium silicate, starch	Friable	Damage, cutting, deterioration
Paper tape	Furnaces, steam valves, flanges, electrical wiring	80	1901–1980?	Polymers, starches, silicates	Friable	Damage, cutting, deterioration Tearing, deterioration
Putty (mudding)	Plumbing joints	20–100	1900–1973	Clay	Friable	Water damage, cutting, deterioration

FIGURE 3.1 Location, composition, and dates of use of asbestos-containing building products.

Abatement strategy	Advantage	Disadvantage	Appropriate application	Inappropriate application	Additional information
Replacement	Permanent solution Allows upgrade Can be integrated with modernization No lead residue left behind on surfaces Low risk of failing to meet clearance standards	Replacement component may be lesser quality than original Replaced components may be high volume and considered hazardous waste Certain installation requires skilled labor	Many interior or exterior components Deteriorated component Highly recommended for windows, doors, and easily removed building components	Restoration projects When historic trust requirements apply Most walls, ceilings, and floors	Nonstandard replacement components may need to be ordered in advance Demolition may damage surfaces May result in increased energy efficiency, for example, if windows are replaced
Encapsulation	Minimal dust if surface preparation is minimal May be faster than some other methods	May not provide long-term protection Requires routine inspection May require routine maintenance Quality installation critical for durability	Exterior trim, walls, floors Interior floors, walls, ceilings, pipes Balustrades	When encapsulant is not appropriate for substrate condition	Encapsulant must be durable and seams must be sealed to prevent escape of lead dust Safe, effective, and aesthetic encapsulants for interior trim components need to be developed and tested Repainting leaded surfaces and using contact paper and paper wall coverings should not be considered for abatement
On-site paint removal	Low level of skill required Allows restoration	Much dust generated Lead residue may remain on substrate and may be difficult to remove Potential difficulty in meeting clearance standards and in protecting workers Stripping agents are hazardous and require more precautions	Limited surface areas When replacement, encapsulation, and off-site removal are impractical	Large surface areas	Unacceptable paint removal methods: – gas-fired open-flame burning – grinding or sanding without HEPA[b] filtration – dry-scraping without misting – uncontained water blasting – open abrasive blasting Chemical removers work best on metal substrates Check with chemical manufacturer regarding recommendations for use on various types of wood and metal substrates
Off-site paint removal	Allows restoration Better finished product generally than with on-site paint stripping	Lead residue may remain on substrate, which may be difficult to remove Damage may occur during removal and reinstallation Swelling of wood, glass breakage, and loss of glues and fillers may occur Hardware left on components may be damaged	Restoration projects, especially doors, mantels, and easily removed thin Metal railings	— — —	Check with stripping company for timing of work and procedures for neutralizing and washing components Check with stripping contractor regarding recommendations for metal substrates

[b]HEPA is high-efficiency particulate air vacuum filtration apparatus.

FIGURE 3.2 Comparison of lead-based paint abatement strategies.

Quality paneling is solid, doesn't warp, and can provide a very attractive finish. Here are several types of paneling:

- Plywood
- Hardboard
- Solid boards

Plywood Paneling

Plywood paneling is the best of the sheet paneling available. It is also an expensive choice. Most plywood paneling comes with three-ply construction. A tough, prefinished face provided the finish. This type of paneling is considered to be the most attractive when it has a veneer of real hard or soft wood.

Hardboard Paneling

Hardboard paneling is much less expensive than plywood paneling, but the hardboard material can warp. The quality of hardboard paneling varies. Some of it is very good, and some of it is not. This type of paneling can be used in moist areas, but it does best when it is applied over drywall. Installing cheap hardboard paneling directly to studs can prove to be a mistake.

Boards

Wooden boards can be used to cover wall studs. These boards can range from weathered barn boards to exotic wood species. It is possible to create some beautiful designs and finished walls with boards, but the cost will be considerably higher than it would be with paneling that comes in 4 x 8 sheets.

PAINT OR PAPER?

Customers sometimes wonder whether to paint or paper their walls. Paint usually wins out, but there is a lot of wallpaper in use, too. It is also not unusual to find a mixture of paint and paper. Borders created with wallpaper between walls and ceilings are quite popular.

> **Don't Do This!** Don't allow yourself to run out of wallpaper before a job is complete. You should make sure that all rolls of paper come from the same manufacturing lot. If you run out and go back for an extra roll, you may find that the individual roll doesn't match the rest of the paper you have already hung.

Don't limit your customers in the range of their imagination. A customer may want wallpaper on the bottom half of a wall with the upper half painted and the two sections joined by a chair rail molding. This can be very popular in a dining room. Or, the customer may want wood wainscoting on the lower half of the wall and wallpaper on the upper half. There is nothing wrong with being creative.

WALLPAPER

Wallpaper is often found in upscale homes and buildings. You may be asked to replace existing wallpaper with new wallpaper (Figure 3.3). When this is the case, it is usually best to remove the old wallpaper. New paper can be installed over old paper in some cases, but I don't favor this procedure. If you opt to layer new paper over old paper, there are a few dos and don'ts to consider:

Dos and Don'ts of Applying New Wallpaper Over Existing Wallpaper

✔ Existing wallpaper must be tight.
✔ Seal existing wallpaper with an oil-based enamel and sizing.
✔ Don't install airtight vinyl wallpaper over existing paper.
✔ Never install new wallpaper over vinyl, foil, flock, or textured paper.

The successful installation of wallpaper requires planning. This planning begins before a job is started and continues throughout the job. Since it assumed that remodeling contractors know how

Paper type	Adhesive type
Regular paper	Wheat or cellulose paste
Lightweight prepasted, paper-backed vinyl	Adhesive installed on back of paper
Lightweight nonpasted vinyl	Powdered vinyl adhesive
Cloth-backed vinyl	Premixed adhesive
Heavy cloth-backed vinyl	Premixed adhesive
Heavy paper-backed vinyl	Premixed adhesive

FIGURE 3.3 Types of wallpaper and adhesives.

to do their job, we won't go into a lengthy discussion of step-by-step instructions for removing old wallpaper or hanging new wallpaper. There simply is not adequate room here for a full disclosure of all the how-to elements of remodeling. Our goal is to make key information accessible to you quickly.

Wallpapering Checklist

✔ Do you have an adequate supply of wallpaper from the same lot?
✔ Do you have the right type of adhesive and enough of it?
✔ Do you have the proper tools on hand?
✔ Have the wall surfaces been prepped?
✔ Do you have a plan for how you will paper the walls?
✔ Have you inspected the wallpaper for defects?

There are several types of wallpaper to choose from. Many of the choices are expensive. Not all types of wallpaper are suitable for all locations. Humidity can be a factor for some types of wallpaper. Be sure to use the right product for the right purpose and in the right place.

- Lining paper is used as a base for some types of wallpaper, such as foils, murals, and burlap. This paper is also very helpful when papering a rough wall. Lining paper is inexpensive

and well worth the investment where you have cracked plaster or other potential problems to overcome.
- Vinyl-coated wallpaper is a standard that can be used for nearly any application, except in rooms where humidity is high.
- Paper-backed vinyl wallpaper is ideal for rooms with high humidity. It is also great for high-traffic areas, like hallways.
- Wet-look vinyl wallpaper is most commonly used in kitchens, bathrooms, laundry rooms, and similar situations.
- Flocked wallpaper is considered a formal wallcovering.
- Foil wallpaper is easy to clean, so it is well suited to kitchens, bathrooms, and laundry areas.
- Burlap or grass cloth can give you a great look, but it does not do well in high traffic areas or where it is in contact with grease, dirt, or high humidity.

Applying wallpaper over latex paint is not a good idea. You are unlikely to get suitable adhesion, and don't apply wallpaper over glossy surfaces or wall surfaces that are peeling or in poor condition. Wallpaper that is applied to a wall that has been covered with a coat of primer is much easier to remove than wallpaper that is installed on an unprimed wall.

PAINT

Paint is, by far, the most common wall finish. Walls and ceilings are painted frequently. Choosing paint colors can be a daunting task. There are so many colors to choose from that anyone can become perplexed. I've always found an off-white or eggshell color popular for walls and ceilings. Of course, it will be your customers who pick the colors of their choice, so this isn't a big factor for a contractor.

Aside from color, someone must decide what type of paint will be used. This can be kept in two categories. You can choose a

 Don't Do This! Don't attempt to thin a textured paint. In doing so, you defeat the purpose of the textured product.

water-based paint, like latex, or you can choose a solvent-thinned paint. Latex is certainly the leader.

Once a decision has been made on the type of paint to be used, you will have to decide how much gloss you, or your customer, wants. I have often used high-gloss products in kitchens and bathrooms. The high-gloss paint is durable for scrubbing. Generally, I recommend a low-gloss or flat paint for general living space. Using a low-gloss paint tends to hide minor defects in wall finishing.

Another key consideration is the type of primer that will be applied before the first coat of paint goes on a wall or ceiling. In some cases, you may also need to consider a stain blocker to make sure dark spots on walls or ceilings don't bleed through the new paint job.

A primer is not always required when repainting a wall or ceiling. However, if you are painting new drywall, changing paint colors, or applying latex paint to a glossy surface, a primer should be used.

Characteristics of Latex Paint

- ✔ Latex paint is water based.
- ✔ Clean up is easy with soap and water.
- ✔ Odors are not offensive in the way that odors from oil-based paint can be.
- ✔ New wall and ceiling surfaces should be primed with a latex or alkyd primer
- ✔ Latex is not for use over unprimed wood.
- ✔ Latex is not for use over metal.
- ✔ Latex is not for use over wallpaper.
- ✔ A full spectrum of gloss finishes is available.
- ✔ Latex dries quickly and may allow two coats to be applied in the same day.
- ✔ Latex adheres well to most surfaces.
- ✔ The durability of latex is not as good as that of an alkyd paint.

It's hard to go wrong with latex paint when working with interior walls and ceilings. This type of paint is so popular that many average people never consider any other type of paint.

Characteristics of Alkyd Paint

✔ Alkyd is a solvent-thinner paint and should not be cleaned up with water.
✔ An alkyd primer is required for raw surfaces when alkyd paint is used.
✔ Alkyd paint dries slower than latex paint.
✔ Odors from alkyd paint are more offensive than those associated with latex paint.
✔ Alkyd paint provides a strong, durable finish.
✔ Alkyd paint is a good choice if you need to hide defects.
✔ Alkyd paint has a synthetic-resin formulation.

Alkyd paint is basically the modern-day replacement of oil-based paint. Since oil-based paint is difficult to clean up, produces offensive odors, and is flammable, it is not high on the list of most painters when it comes to preferred paint.

Characteristics of Urethane and Polyurethane

✔ These products can be used over nearly any porous surface or existing finish.
✔ These products are usually solvent-thinned.
✔ The finish from these products is basically clear.
✔ These products resist grease, abrasion, and dirt.
✔ These products are often applied to wood trim.

Epoxy paint is not normally used in everyday remodeling. This type of paint will not adhere to previously painted surfaces. Epoxy can be applied to unpainted, nonporous surfaces, but a hardening agent is often needed just before application, and the mixing process can be tricky. Under normal conditions, there is little reason to use this type of paint.

Characteristics of Textured Paint

✔ Textured paint is often used on ceilings.
✔ Some textured paint comes premixed.

✔ Minor defects in wall or ceiling finishing can be hidden with textured paint.
✔ You may have to add a sand-like powder to create textured paint with some brands.
✔ Do not thin textured paint.

Characteristics of Acoustic Paint

✔ Acoustic paint is used to paint acoustic ceiling tiles.
✔ Acoustic paint does not affect the sound-deadening qualities of acoustic tiles.
✔ One coat of acoustic paint should be all that is needed.
✔ A primer is not needed with acoustic paint.
✔ Acoustic paint can be applied with a sprayer or a special roller.
✔ Water is all that is needed to clean up acoustic paint.

TILE

Ceramic tile is not as common on walls as it once was, but it is still appreciated in kitchens and bathrooms. Most bathrooms these days have painted walls and fiberglass tub surrounds. It was not always this way. As a remodeler, I am sure you have done your share of rip-outs where ceramic tile was torn off walls and hauled away. Should you consider tile a dead issue for walls in modern remodeling? I don't think so.

Many homeowners think of quality when they think of tile tub surrounds. I wouldn't begin to tell you that customers will pick tile over fiberglass in a major way, but there is still a market for tile tub surrounds. Don't overlook this option that can give your jobs a distinctive look.

Kitchens are a great place to use tile for wall finishing. Tile is so easy to clean that it is a natural for kitchen walls around cooking surfaces and counters. I'm not saying that every wall in the kitchen should be tiled, but there are definitely viable places to suggest tile to your customers during kitchen remodeling (Figure 3.4). Many of my past customers have been thrilled with the designs and images created with tile.

Tile	Sizes
Ceramic tile	Sizes range from 1-inch squares to 12-inch squares Most tiles have a thickness of 5/16 inch
Ceramic mosaic tile	Sizes range from 1-inch squares to 2-inch squares Also available in rectangular shapes, generally 1"-x-2" Average thickness is 1/4 inch
Quarry tile	Sizes range from 6-inch squares to 8-inch squares Rectangular tile is available in 4"-x-8" size Typical thickness is 1/2 inch

FIGURE 3.4 Tile sizes.

Trade Tip: Have your tile distributor or subcontractor provide you with some sample tile boards that you can carry with you when bidding jobs. Showing customers a selection of ceramic tile can be all it takes to set you apart from your competitors.

FLOORING

When you get into flooring, there are many possibilities. The routine is carpeting throughout most living space, with vinyl in kitchens and bathrooms. This is so standard that many builders, remodelers, and property owners don't think much further about it. To give your jobs a distinctive edge, consider incorporating some quarry tile in entry ways, some ceramic tile in bathrooms, finished wood flooring in formal areas, and some more quarry tile in kitchens. This does run the cost of a job up, but the work stands out and many customers are willing to pay for it if you show them why they should.

CARPET

Carpet is the king of flooring. It is used in office buildings, homes, schools, and a host of other types of buildings. The question is, what is the right carpet for the application? How important is the

Don't Do This! If it hasn't happened to you yet, it will. Some customer is going to ask you to carpet their bathroom. They are going to want a warm fuzzy floor to walk on. I strongly suggest that you discourage this. Most code requirements call for an impervious floor in bathrooms. This is done for a reason. Carpet is next to impossible to clean when used in a bathroom, and the water that is bound to get on the carpet can cause odors, mildew, and potential wood rot on the floor below the carpet pad.

type and quality of the carpet pad? What makes one carpet type cost three times what another type of carpet costs? Questions like these are common when choosing the proper carpet (Figure 3.5, 3.6).

The choice of carpeting is usually made by the customer. You can refer to the information in Figure 3.5 for features of the various types of carpet. If you have a good relationship with your carpet installer or supplier, you can let your customers talk directly to the experts. Personally, I like to attend such meetings to make sure that my suppliers and subcontractors give my customers the right information in the right way.

CARPET PADDING

Carpet padding is very important. If you have to choose between an inexpensive carpet or an inexpensive pad, take the inexpensive carpet. This point was shown to me over 20 years ago in a demonstration that is still fresh in my mind. At the time of the demonstration, I was building 60 homes a year while

Did You Know: that felt carpet padding is more durable than foam? It is. However, don't install a felt pad in an area where high humidity is expected.

Carpet	Qualities
Polyester	Bright colors, resists mildew and moisture, stays clean
Olefin	Very durable, resists mildew and moisture, very stain-resistant
Wool	Durable, abrasion-resistant, reasonably easy to clean, should be protected against moths, resists abrasion.
Acrylic	Resists mildew, resists insects, remains clean, resists abrasion
Nylon	Extremely durable, resists abrasion, resists mildew, resists moths, remains clean, tends to create static electricity

FIGURE 3.5 Carpet features.

Carpet	Cost
Polyester	Moderate cost
Olefin	Prices vary
Wool	Expensive
Acrylic	Moderate cost
Nylon	Prices vary
Polyester	Moderate cost

FIGURE 3.6 Cost comparison of carpet fibers.

running the remodeling and plumbing divisions of my business. Customers used to make their own selections of carpeting and padding for remodeling jobs, but I often chose the products for the houses we were building on speculation. This is where the lesson came into play.

A salesman in a carpet store showed me three pieces of carpet padding. They all looked similar, but they were not the same price. The salesman put the three pad samples on the floor. Then he handed me a cheap carpet sample to put on the padding. I was wearing work boots with lug soles. The salesman handed me a stopwatch and told me to hold onto it. Then he told me to stand on the carpet for 30 seconds. I did. Then I stepped off of the carpet and was told to time how long it took for my foot prints to go away. Frankly, I don't recall the exact amount of time that it took for my footprints to disappear. But, this was the cheap pad and it took a fair amount of time.

Once my footprints were gone from the first experiment, I was told to move the carpet sample to the second piece of padding and repeat the previous steps. I did, and the footprints went away quicker.

When I got to the most expensive piece of padding, my footprints were gone very quickly. The point being made was that the pad has a lot to do with the wear on the carpet fibers. Then

> **Don't Do This!** More and more floors are being built with radiant heat concealed in the flooring. Don't use a rubber-backed carpet on this type of flooring.

we used an expensive carpet sample to repeat the test. The trick was in the padding. This is when I learned that you can use a less expensive carpeting if you use a high-quality padding.

VINYL FLOORING

Sheet vinyl flooring is extremely common in foyers, kitchens, laundry rooms, and bathrooms. It is an affordable, durable, and dependable flooring. One key to keep in mind with sheet vinyl is having a smooth, solid base beneath it. You can install new vinyl over old vinyl, but make sure that the existing flooring is tight and smooth if you chose to do this. I recommend removing the existing vinyl, prepping the subflooring, and installing the new vinyl over, adding the new flooring on top of the existing flooring.

Characteristics of Inlaid Sheet Vinyl Flooring

- ✔ Inlaid vinyl flooring offers advantages over other types of vinyl flooring.
- ✔ There are many patterns and surfaces available for inlaid vinyl flooring.
- ✔ Inlaid vinyl is easy to maintain.
- ✔ There are few seams associated with inlaid vinyl flooring.
- ✔ Inlaid vinyl is thick and heavy and feels good underfoot.
- ✔ Cost can be prohibitive for inlaid vinyl flooring.
- ✔ The durability of inlaid vinyl is tough to beat.

There are three primary choices when it comes to sheet vinyl flooring. Inlaid vinyl is the cream of the crop, but it is expensive. Rotovinyl is a good alternative to inlaid vinyl. Stretch, cushioned vinyl is newer than the other types of sheet vinyl. It also offers some nifty features.

Characteristics of Rotovinyls

✔ Rotovinyls are thinner and lighter than inlaid vinyl. This makes them easier to install.
✔ Designs are printed on the surface of rotovinyls.
✔ Rotovinyls are comfortable to walk on.
✔ Sound-dampening is an advantage associated with rotovinyls.

Characteristics of Stretch Cushioned Vinyl

✔ Stretch cushioned vinyl is light in weight and cushioned for a comfortable feel.
✔ The elasticity of stretch cushioned vinyl allows it to contract after installation to prevent wrinkles and bubbles.
✔ Minor imperfections in subflooring are not a problem when stretch cushioned vinyl is installed.

Sheet vinyl flooring is the hands-down winner for average remodeling jobs that involve kitchens, bathrooms, and laundry rooms. But, if your customer wants something special, consider offering them a nice tile floor.

TILE FLOORS

Tile floors used to be standard procedure in bathrooms. This is not the case in today's market. Much of the reason is cost. Another reason is that a ceramic tile floor can be quite cold and slippery. Many builders and remodelers are using radiant floor heating in cold climates. When this is done, a tile floor is great for being a heat sink. There is a place for tile floors in modern homes.

I love quarry tile in mudrooms and foyers. Personally, I can't think of a better choice for these locations. A tile floor is appropriate for a laundry room, a kitchen, a sunroom, and other special-use locations. Textured tiles are fantastic when it comes to avoiding a slippery surface. There are so many types, styles and designs to choose from when it comes to tile that it can be per-

Did You Know: that you can install radiant, in-floor heating without the use of a concrete floor? You can, and it is well worth considering for customers who heat their homes with hot-water heat. This type of heat is becoming extremely popular in cold climates.

plexing to settle in on a single choice. From small squares of ceramic tile to large 12 inch squares of quarry tile, with other sizes in between, there will be no shortage of options for customers who are interested in the quality of a tile floor.

WOOD FLOORING

Wood flooring used to be found in countless living rooms and dining rooms. You don't see it often anymore. Why? I suspect the cost is a major reason. Care and maintenance are probably other reasons why wood is not as popular as it once was. With working parents, busy lifestyles, and similar changes in the way of life, the lure of wood flooring is not as strong as it once was.

I remember being a child who used to run the buffer on the hardwood floors in the hall, dining room, and living room of our family home. It was fun, and I got an allowance for doing it. Somehow I doubt that a lot of children today would find running a buffer fun. In any case, wood flooring is distinctive and gives a certain feel to a room that no other type of flooring can.

Three Types

There are three basic types of wood flooring. There is strip flooring that is comprised on narrow, random strips. Wide planks are

Trade Tip: When you take delivery of wood flooring, be sure to store it in a heated space. The temperature range should be 65-70°F. Storing wood flooring in an unheated area, in a cold climate, can result in warped wood.

another type of wood flooring. Then there are small, wood squares that are used to create a finished wood floor. The thin strips were very popular for many years in formal situations. Wide planks are favored for rustic homes. The wood tiles have a good reputation and are used for multiple purposes, but I have never had a lot of customers who found the squares desirable (Figures 3.7, 3.8).

Once you, or your customer, decides on a floor style, you have to decide on a floor type. Will the wood be hardwood or softwood? Narrow, strip hardwood flooring is very popular in formal settings. Wide, softwood planks play well in recreation rooms, family rooms, and rustic homes. Wood tiles can be used in foyers, kitchens, or any other room where they are wanted.

STRIP FLOORING

Strip flooring is extremely popular. This type of flooring is installed with a special nail gun. The nails are installed at a sharp angle so that they are not seen once the flooring is installed. This

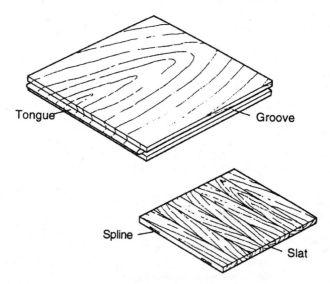

FIGURE 3.7 Two types of wood block flooring.

type of flooring is of the tongue-and-groove type (Figures 3.9, 3.10). Most strip flooring is made of oak, but the flooring is available in a variety of wood choices. Care must be taken in the planning of an installation to arrive at a pattern that is acceptable to you and your customer.

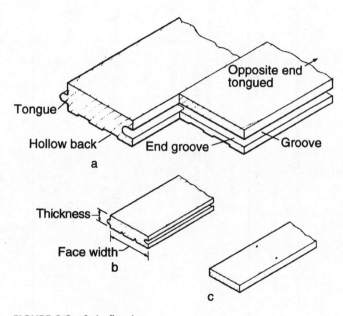

FIGURE 3.8 Strip flooring.

 Don't Do This! When you are discussing wood flooring with customers, don't refer to the flooring as hardwood flooring unless it is hardwood flooring. I know of real estate brokers who have been sued for representing a wood floor as a hardwood floor when the flooring was actually made from a softwood product (Table 3.1).

TABLE 3.1 Grades and Descriptions of Strip Flooring

Species	Grain orientation	Size (in.)		First grade	Second grade	Third grade
		Thickness	Face width			
Softwoods						
Douglas-fir and hemlock	Edge grain	25/32	2-3/8–5-3/16	B and better	C	D
	Flat grain	25/32	2-3/8–5-3/16	C and better	D	– –
Southern Pine	Edge grain	5/16–	1-3/4 by 5-7/16	B and better	C and better	D (and no. 2)
	Flat grain	1-5/16				
Hardwoods						
Oak	Quarter sawn	3/4	1-1/2–3-1/4	Clear	Select	No. 1 common
	Flat grain	3/8	1-1/2, 2			
		1-1/2, 2				
Beech, birch, maple, and pecan		3/4	1-1/2 by 3-1/4	First grade	Second grade	Third grade
		3/8	1-1/2, 2			
		1/2	1-1/2, 2			

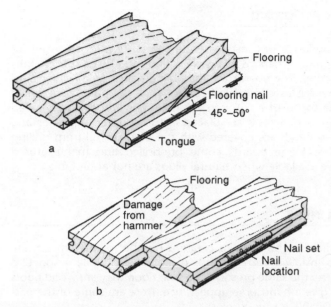

FIGURE 3.9 Nailing procedure for strip flooring.

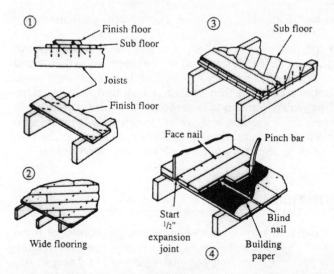

FIGURE 3.10 Nailing detail for tongue-and-groove strip flooring.

PLANK FLOORING

Plank flooring is not installed with the same methods used for strip flooring. When installing plank flooring, the boards are butted together. The boards are then blind nailed, in a fashion similar to what is used on strip flooring. Then the boards are drilled from the top with a countersunk hole. The wood plug is saved for future use. Screws are placed in the countersunk holes and the screw heads are covered with the plugs saved from drilling the holes. Some boards come pre-drilled and manufactured plugs are available when natural plugs are not acceptable.

WOOD TILES

Wood tiles can be used to create a number of patterns. These tiles are installed with the use of adhesives. While wood tiles don't carry the same prestige of a strip floor, they are wood floors and do give a handsome appearance. Here are some of the patterns available for wood tiles:

- Brick
- Finger
- Finger laid diagonally
- Herringbone
- Foursquare
- Foursquare laid diagonally

Once the flooring is installed, you will need to install finish trim (Figure 3.11). The installation of trim requires patience and precision. There is no mystery to installing trim. Even so, a great many contractors install trim in such a way that few reputable contractors would want to take responsibility for it. I've been in the business since the 70s and I've only seen a few trim carpenters who did their jobs at a level I would find acceptable. The trim is there for all to see. It is the icing on the cake. Don't blow it!

INTERIOR TRIM

Interior baseboard trim is usually considered to consist of three styles. There is clamshell, colonial, and strip molding (Figure 3.12). My personal favorite is colonial. Clamshell molding is usually associated with cheap construction. Strip molding is fine in a

Flooring and Wallcoverings

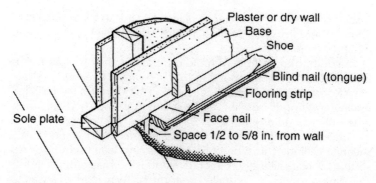

FIGURE 3.11 Trim installation for strip flooring.

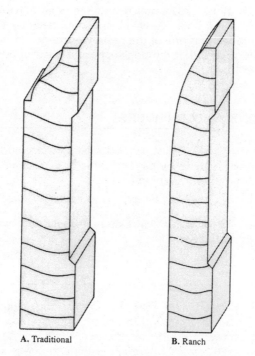

FIGURE 3.12 Types of baseboard moldings.

rustic home, but it looks out of place in many styles of buildings. Some other terms for molding include traditional baseboard and ranch baseboard.

In addition to baseboard molding, there is shoe molding. This is small, quarter-round molding that is typically used when sheet vinyl flooring is installed. Shoe molding is not normally installed when carpeting is installed. Shoe molding is also used, at times, when tile is used as a finish flooring.

Crown molding is very decorative and installed where a wall meets a ceiling. This type of molding is normally used in formal settings. Chair rail is also decorative and is normally found in formal dining rooms. It is installed near the halfway mark of a wall to keep chairs from dinging the wall. Chair rail is commonly used to join two types of wall finishes, such as a wainscoted bottom half of a wall and a painted or papered upper half of the same wall.

Well, let's move to Chapter 4 and see what can be learned about kitchen remodeling. Of all the remodeling work done, kitchen remodeling is one of the best investments for homeowners, and this means that remodelers have a great opportunity to specialize in the field.

QUESTIONS AND ANSWERS

Q: I've heard that it is not a good idea to use wood floors in rooms that will experience a lot of traffic or a lot of activity from children. What is your opinion on this?

A: *If the wood is sealed with a protective coating, it should be fine.*

Q: Is tile too fragile to install in a kitchen where pots or pans may be dropped on the floor?

Don't Do This! Don't use finger-joint trim if you are planning to stain it. The finger-joints will show through the stain. Believe it or not, I have seen homes where finger-joint trim was installed, stained, and left for all to see. This is horrible. Finger-joint trim is cheaper than solid trim, and it is fine if you paint it, but don't stain it.

A: *Tile can be cracked or broken when hard objects are dropped on it. However, this usually isn't a big problem.*

Q: I just finished a job and the vinyl flooring that my subcontractor is installed has a bubble in it. According to the sub, he can puncture the bubble and repair the floor. I think the homeowner should get a new floor. What would you do?

A: *This is a tough situation. It is feasible to pop bubbles and seal them with good results. Another option would be to cut out a section of the floor and patch it. Getting a new floor is the best option, but the subcontractor will probably be very resistive on the issue. Since you are responsible to the customer, I recommend that you see what will satisfy the customer and have the customer give you written authorization for whatever decision is made. Then make your sub do what is necessary to make the customer happy.*

Q: During the rough finish of walls in an addition I'm doing, the customer's child drew on the wall. Is this safe to paint over?

A: *Talk to your painter. I suspect that you will be told to apply a stain blocker over the marks on the wall prior to painting the wall.*

Q: I've got to remove old wallpaper and replace it. Do those steamers that you can rent work?

A: *They do work, but if the wallpaper was installed on drywall that was not primed or painted, the removal process can be very frustrating.*

Q: Can I use an oil-based paint in the winter for interior use?

A: *You can, but there are issues to consider. The paint may not dry quickly. Some people don't deal well with the odors from oil paint. Since opening windows in the winter for ventilation creates problems, you might find that your customers are offended by the odors.*

Chapter 4

KITCHEN REMODELING

Kitchen remodeling may well be one of the most profitable elements of residential remodeling. It is widely documented that kitchen remodeling offers some of the highest rates of return on a homeowner's investment. This creates an excellent opportunity for remodeling contractors.

Many contractors specialize in kitchen and bath remodeling. When I lived in Virginia, I had an entire division devoted to kitchen and bath remodeling. If you live in a populated area, you should be able to focus on only kitchens and baths and do quite well for yourself.

There are a number of reasons why kitchens are lucrative. Cabinets can be very expensive, and they often hold generous mark-ups for contractors. If I had to pick one element of kitchen remodeling that makes it a cash cow, I would say it is the cabinetry.

It is no secret why kitchens are popular remodeling projects. A kitchen is one of the most used rooms in a house. Since people often associate cleanliness with food, it makes sense that kitchens are a subject of interest when it comes to keeping up appearances.

It is common in older homes to find small kitchens. Many of the kitchens are not bright and cheerful. This opens the door to remodeling. Adding skylights, garden windows, new cabinets and counters, and other enhancements is always popular.

Expanding a small kitchen is sometimes possible without major structural work. This, of course, robs some other portion of a home of some space, but it may still be desirable. When you can't expand a kitchen, you may be able to make it appear more spacious with lighting and wall modifications. Creating a desired effect can be accomplished in many ways, so use your creative juices when you are evaluating a kitchen job.

RIPPING OUT

Ripping out an existing kitchen is a lot of work. While many homes have more than one bathroom, they rarely have more than one kitchen. Ripping a kitchen out creates a serious disruption for most families. You have to keep this in mind. Are you going to keep the kitchen sink, range, and refrigerator functional until the last minute? If you are, your workers won't be able to work as freely or as quickly as they might like. Factor this in when you bid a kitchen job.

Do the appliances need to be replaced? Most appliances are replaced during a major remodeling effort. Physical appearance is often a motivator in replacing appliances. However, the age of appliances and their estimated useful life should be considered for appliances that are still attractive. You may not have to replace the appliances (Figures 4.1, 4.2).

Rip-outs are always dirty and dusty. Before you rip out cabinets, you need to know for sure if you are going to replace them, reface them, or reinstall them. Most of the rip-out work is fairly standard stuff. I think the two keys here are the cabinets and keeping the kitchen functional for as long as possible.

Appliance	Average life expectancy (years)
Range	15–16
Refrigerator	13
Dishwasher	11
Garbage disposal	10

FIGURE 4.1 Useful life of appliances.

Number of bedrooms	Refrigerator size (ft^3)
2	14
3	17
4 and 5	19

FIGURE 4.2 Recommending sizing of refrigerators.

KITCHEN EXPANSION

Kitchen expansion is often wanted when a kitchen is being remodeled. Frankly, most kitchen remodeling doesn't involve expansion. This is due to cost and space limitations, but most homeowners would like to have a larger kitchen. What can you offer your customers when it comes to expansion?

Interior expansion is cost effective. If you have rooms adjoining a kitchen area where you can steal some space, this is a great option. It doesn't take a lot of extra space to make a big difference in the perceived space of a kitchen. If you can sneak two feet out of an adjoining room and put it in a new kitchen, the results can be dramatic.

Unfortunately, many homes don't have abundant space adjoining their kitchens. Where does this leave you? If you can't borrow space from some other room, you have to consider adding space. This tends to be expensive. Will your customer be willing to add onto an existing kitchen? Is there enough space outside of the home to allow an addition without violating zoning set-back requirements? This is certainly something that you have to confirm before you agree to build an addition.

A compromise can come in reconfiguring existing space (Figure 4.3). This might mean tearing out walls and reframing them with attributes to give more perceived space. For example, if you have a solid partition wall, you might cut a large section out of it to install a pass-through opening and fit it with folding doors that can be opened and closed. Changing traffic patterns and

Don't Do This! Don't expand the foundation of a home until you verify that there is adequate space for the addition without violating local zoning requirements. I know of some commercial buildings that were built in violation of zoning set-backs. Believe me, the contractors were not happy campers.

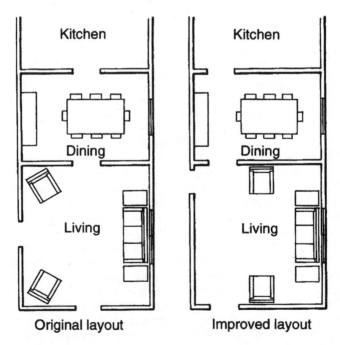

FIGURE 4.3 Example of redesigning traffic patterns for more user-friendly space.

door locations can improve the use of kitchen space. You have to look at all options.

Another option is redesigning an existing kitchen. This can be fairly easy to accomplish. There are limitations to what you can do with existing space, but the work triangle of a kitchen can be improved in most cases. You can often do this without expanding the footprint of a kitchen. It can be amazing what a new design layout will do for the functional use of existing space.

A U-shaped kitchen keeps the work area centered around one end of the room (Figure 4.4). Distance between the kitchen sink and appliances vary in these types of kitchens, but the U-shape design opens up an option for an eat-in kitchen

Kitchens laid out with an L-shape offer good flexibility for the use of floor space (Figure 4.5). Given enough space, this style of kitchen gives the user room to roam, but a larger kitchen is usually needed to make this viable for an eat-in kitchen. You must

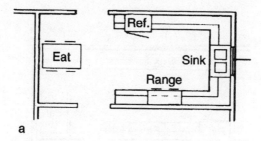

FIGURE 4.4 A U-shaped kitchen.

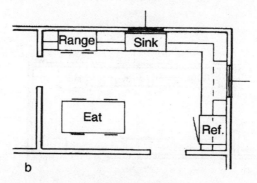

FIGURE 4.5 An L-shaped kitchen.

consider traffic patterns when choosing a design. One potential problem with an L-shaped kitchen is the distance between work spaces and appliances.

Galley or corridor kitchens tend to be small. The joke in the building business is that you have to move sideways through a galley kitchen. If you have limited space to work with, a galley kitchen may be your only viable option (Figure 4.6). As small as a galley kitchen may be, it can give you a good work triangle. It's even possible to place a small eat-in table at one end of a galley kitchen.

Sidewall kitchens are not much different than galley kitchens. The main difference is that all workspace is placed on one wall. A breakfast bar can be incorporated into a sidewall kitchen, but a table is usually not feasible. This is the smallest of the small when it comes to kitchen layouts (Figure 4.7).

Country kitchens that open into a family room give a lot of space (Figure 4.8). It's important that the customer doesn't mind

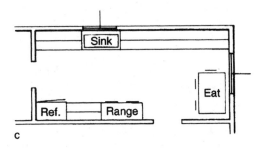

FIGURE 4.6 A galley kitchen.

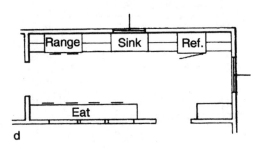

FIGURE 4.7 A sidewall kitchen.

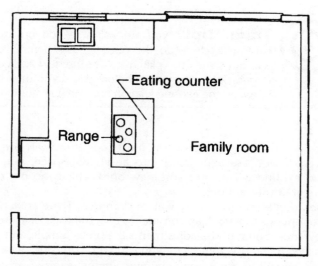

FIGURE 4.8 Country kitchens.

people sitting in the family room and looking into the kitchen, but this design is easy to make happen. Essentially, you demo the wall between the kitchen and the living room or family room and make it flow together. This style of kitchen is very popular and extremely effective.

Expanding a kitchen is usually thought of as adding square footage. Making an open kitchen, like a country kitchen, doesn't require additional square footage. It is accomplished by design methods. If you have to add square footage, you have to use existing space from some adjoining room or add onto the home. A lot of houses don't have enough space to spare in adjoining rooms. This limits you to full expansion. At this point, you are getting into a room addition. Refer to Chapter 6 for details on room additions.

MANIPULATING WALLS

Manipulating walls in kitchen remodeling is not usually a complex issue. This is very basic remodeling with what are normally not load-bearing walls. Generally speaking, this is just a matter

> **Trade Tip:** If you are going to add onto an existing kitchen, do the outside work before you tear out the existing kitchen. Get the new addition under roof and sealed in before you demo the wall between the existing kitchen and the new addition.

of building simple partitions. In most jobs, the walls are not moved. If you are expanding into additional space, you will demo existing walls and frame new ones. This is no big deal. There are other options.

You don't have to move a wall to change it. There are a number of options that you can consider to give the walls of a kitchen a new look. Some of the adjustments are pretty simple, but very effective. If you use your head, you can come up with ideas that your competitors are not offering customers.

A Few Tips for Wall Modifications

- ✔ Cut in a pass-through window to a dining area.
- ✔ Create a framed area for decorative display. For example, install a piece of fixed glass between the kitchen and another room that has a theme to it. I once did a kitchen with a loon theme. It breaks up the wall structure and adds character to the room. The fixed-glass inserts were done with color and etchings of loons.
- ✔ Cut fixed-glass panels in above wall cabinets. It is often difficult to reduce wall space in small kitchens, but if you go above the cabinets, you don't have a problem. The glass panels add additional lighting to the kitchen, and they add flair to the design.
- ✔ Consider putting in a half wall that will accommodate base cabinets and a counter and then have spindles run to the ceiling. This allows workspace without boxing in the kitchen.
- ✔ If you are dealing with an exterior wall, replace a standard kitchen window with a garden window.
- ✔ Install a hexagon window that opens and has a screen. It will produce additional light and a great look for the kitchen.

KITCHEN FLOORS

Kitchen floors are often covered with sheet vinyl. This is a good choice. It is economical, easy to clean, durable, and hard to beat. As a builder and remodeler, this has been the flooring that I have used most often. As common and intelligent as this choice is, it is not the only choice.

Personally, I favor tile flooring in kitchens. I love quarry tile for kitchen floors. This is just my personal preference. But, a lot of homeowners have agreed with me over the years. Sheet vinyl probably makes the most sense for a kitchen floor, but tile is special. And, you can do some great designs with it. It costs more and plates dropped on it are more likely to break. Some types of tile become slippery, and this has to be avoided. Textured tile provides good footing and a good look. For high-end jobs, I really like tile.

Wood flooring is a potential option for kitchens. It is very appropriate for a country kitchen that has a rustic venue. The wood can be hardwood or softwood. My experience shows that softwood is more popular. Wide boards are the typical wood flooring for kitchens. I have no problem with this, but I have had few customers who chose this type of flooring. By in large, vinyl is the winner hands down, with tile coming in second. Carpet is sometimes installed, but I have not seen much of this. I suppose it is too difficult to clean.

KITCHEN WALLS

Kitchen walls can be painted, papered, tiled, or treated with other expressive interests. Most kitchen wall are painted with a high-gloss paint. The high-gloss paint is used to allow for periodic scrubbing. If you choose to install wallpaper, use one that is not affected by humidity. Kitchens can be hard on wallpaper.

 Trade Tip: If you are going to replace a vinyl floor with a wood floor, the height of the floor will be higher. This creates a tripping risk and the need for a transition threshold to the room.

Stenciled patterns where walls meet ceilings and wallpaper stenciling for the same purpose are very popular. The basic rule for kitchens is to avoid raised surfaces. Keep surfaces smooth and easy to clean. This will keep you out of trouble.

CEILINGS

Textured ceilings are usually not a good idea in kitchens. There is a lot of grease released in kitchens. When a textured ceiling is in place, it is a dust magnet. Most people don't like this. With rare exceptions, such as a textured ceiling with exposed beams, a textured ceiling is not a good idea in a kitchen. Use a painted ceiling that can be cleaned.

ELECTRICAL WORK

Electrical work in kitchens should often be updated. Many homes that are remodeled are older homes. This can be more expensive than you might think. Can the existing power panel handle modern electrical needs? Your kitchen remodeling might be covered under some grandfather clauses for electrical requirements. If you are doing a major job, the modern codes will probably have to be met. This is something that you have to think of, plan for, and budget for.

If you start a job and then find that an existing electrical service has to be upgraded to meet code requirements when you have not factored in this cost, you can lose a lot of money. Check it out! Have your electrical contractor advise you on electrical needs prior to contracting the job.

Did You Know: that modern electrical codes require major kitchen appliances to have their own circuits? This was not always the case, but it is a provision in the modern code. Check with your local code officer to see if your job is grandfathered or if it will need major rewiring.

The panel box and service is not the only question. Older homes had shared circuits. New codes require many separate circuits. How hard will this new wiring be to install? Bottom line: pay attention in the bidding stage and avoid costly mistakes.

PLUMBING

Plumbing in a kitchen is usually not very complicated. There is, of course, a kitchen sink that requires plumbing. Many kitchens are equipped with dishwashers that require a water supply and a drain. It is quite likely that a refrigerator will require a water supply for an icemaker. A garbage disposer is a common plumbing appliance in kitchens. There can be other plumbing fixtures, appliances, and piping found in a kitchen. All in all, there is not a lot of plumbing to be dealt with when remodeling a kitchen.

How much plumbing work is required in a kitchen remodeling job depends on the extent of the remodeling plans. If all you are doing is removing and replacing appliances and fixtures in their existing location, the work required is minimal. However, if you are moving walls, relocating fixtures and appliances, or eliminating walls, the work can be fairly extensive.

Water supplies to plumbing fixtures and appliances should be equipped with shutoff valves. Not all water supplies have these valves. Sometimes the valves that exist don't function properly. Under normal conditions, the valves can be turned off and the removal of plumbing connections to appliances and fixtures is easy. Care must be taken to avoid crimping or breaking water supplies during the removal phase.

Installing new plumbing appliances and fixtures in existing locations is not a complicated process. Even so, there may be some need for modifications. If you are working with an older home where galvanized pipe was used as a kitchen drain, you should replace as much of the drain as possible. This type of drain rusts and holds grease. It's common for galvanized pipe to gradually build up residue until there is no opening in the drain. A plumber can snake the drain and make a hole in the residue, but this is only a quick fix. In time, the drain will close up. I strongly suggest that you replace as much galvanized drainpipe as you can with plastic drainpipe.

When walls are removed or relocated, you may have to move existing plumbing pipes. Water pipes, drains, and vents may

Don't Do This! Don't connect new plumbing drains to existing galvanized drains when you can replace the galvanized pipe. If you make a connection from a kitchen drain to a galvanized pipe, you are likely to have problems with drain stoppages. This is especially true if a garbage disposer is installed where one did not exist before.

have to be moved. This can get expensive, so you should pay attention to probable plumbing locations when you are pricing a job. The relocation of these pipes is not such a big deal in a one-story home that has an open basement or crawlspace, but it can be very tricky in multi-story homes, houses built on concrete slabs, and homes where the area beneath the kitchen floor is finished space.

Sinks

Kitchen sinks come in many shapes, sizes, colors, and material types. Stainless steel is the most common type of kitchen sink. These sinks may have a single bowl, two bowls, or more than two bowls. A double-bowl sink is the most common.

When choosing a stainless steel sink, don't buy the cheapest one you can find. The quality of the sink does matter. Light gauge

Trade Tip: If you are inspecting a job to offer a bid to remodel a kitchen where walls will be moved or removed, look for evidence of plumbing in the walls that will be affected. If you can get under the kitchen floor and see the floor joists, you can tell where plumbing pipes extend upward. Go outside and look for vent pipe locations as they terminate the roof. If there is attic space over the kitchen, look in it for vent pipes. Vent pipes are often offset, so the terminal location at the roof is not always indicative of where the pipe comes up through the kitchen.

sinks can bend when a garbage disposer is attached. Invest in a decent sink to avoid customer complaints.

Cast-iron sinks are not used very often these days. There is still a market for them, but it is usually in upscale housing. These sinks are heavy and expensive. However, they offer many designs and colors to choose from, and this makes them popular with some homeowners. As you talk with your customers about sinks, you will find that there are a multitude of choices for them to choose from.

Faucets

When it comes to kitchen faucets, you can find dozens upon dozens of styles, types, designs, and colors. In the simple stage, you start with choosing between a single-handle faucet or a two-handle faucet. Then you will have to decide if the faucet will be equipped with a spray or any other accessory. From here, it gets crazy with options.

Your customer might want an enameled faucet in a bright color. It could be that a gold faucet is desired. Most customers will go with a standard chrome finish. The bottom line here is that you should expect to spend some time with your customers in choosing a suitable faucet.

HEATING

Heating is normally only a problem in kitchen work when walls are moved. If the basic footprint of the kitchen remains intact, you should not encounter any complicated heating tasks. Like plumbing, heating may have to be relocated if walls are moved. This can be as simple as extending heating ducts or heat pipes, or it can be much more complicated if a ducted return occupies a wall that is being removed. If you are going to demo a wall, check to see what might be in it.

WINDOWS AND SKYLIGHTS

Windows and skylights can enhance a kitchen dramatically. Most people prefer light, cheerful kitchens. Many older kitchens are dark and dreary. Replacing an existing window with a garden window

Did You Know: that heat ducts and pipes are sometimes installed within kitchen cabinets? You may discover this to be the case, so check for any problems you may encounter with cabinet removal when heating components are located in the cabinets.

makes a dramatic statement. When skylights can be installed, they provide superb natural lighting. The expense of quality skylights and windows can be substantial, so don't underestimate the cost.

Existing windows can be a problem when you are rearranging a kitchen. Keep this in mind. There are creative ways to get around this type of problem. For example, you could design a wall cabinet layout that incorporates the window as a part of the overall presentation. Pay close attention to existing windows and how they will play into your scheme for a new kitchen design.

APPLIANCES

Appliances are a natural part of a kitchen. At a minimum, you will have a refrigerator and a range. The potential list of kitchen appliances is a long one. If you are taking responsibility for the acquisition and installation of appliances, make sure that you obtain appliances that will fit in the space available. Having a refrigerator that is one inch too wide for the space provided is a serious situation. Make sure you leave adequate space for the appliances your customers want.

CABINETS

Of all the expenses involved in kitchen remodeling, cabinets rule the roost (Figure 4.9). There is big money in cabinets, and there is plenty of profit in them for contractors. Getting customers to make decisions on cabinet selection can be a time-consuming process, but it's worth it. Just getting customers to settle on a type of cabinet can be difficult.

Price is often the yardstick used to measure which type of cabinets to consider (Figure 4.10). People are usually restricted by

Type of cabinet	Features
Steel	Noisy Might rust Poor resale value
Hardwood	Sturdy Durable Easy to maintain Excellent resale value
Hardboard	Sturdy Durable Easy to maintain Good resale value
Particleboard	Sturdy Normally durable Easy to maintain Fair resale value

FIGURE 4.9 Kitchen cabinet features.

Type of cabinet	Price range
Steel	Typically inexpensive, but high-priced units exist
Hardwood	Typically moderately priced, but can be expensive
Hardboard	Typically moderately priced
Particleboard	Typically low in price, but can reach into moderate range

FIGURE 4.10 Kitchen cabinet price ranges.

some sort of budget. Whenever possible, the choice of kitchen cabinets should not be held to the lowest price. Some cabinets are far too expensive for an average homeowner to feel comfortable with. Since cabinets are a key feature in a kitchen, serious consideration should be given to them.

Custom Cabinets

Custom cabinets can be extremely expensive, but they are sometimes competitive with off-the-shelf cabinets. As a builder and remodeler, I have used both stock cabinets and custom cabinets. Some people feel like anything less than custom cabinets is poor in quality. This simply is not true. I have used several brands of stock cabinets and have found them to be of outstanding quality (Figure 4.11). Don't get me wrong, you can buy cheap, bad cabinets, but you can also get shoddy workmanship in custom cabinets.

The one big drawback to custom cabinets can be the time it takes to have them built. Along these same lines, if you gut a kitchen and are counting on custom cabinets to be delivered on a certain day, what are you going to do if they don't show up? I've had this happen and it is not an enjoyable experience. When you work with stock cabinets, you can get them all delivered to your supplier and know when they can be put on your job.

Good cabinetmakers can turn out works of art. They can also create unique pieces. These advantages mean a lot to some customers. Don't rule out custom cabinets, but be aware of the potential time lags and the probable additional cost.

Stock Cabinets

Stock cabinets are readily available. Quality ranges from mediocre to excellent. Don't think that cabinets will be cheap just because they are produced in mass numbers. Many stock cabinets are far from cheap. I have found over the years that stock cabinets provide suitable results for the long-term satisfaction of my customers. The key is to inspect cabinet models closely to ensure quality. Looks are not the only test of suitability for cabinet selection.

When you begin to go over cabinet options with your customers, you will have a lot to talk about. I suggest that you obtain catalogs from a variety of quality cabinet manufacturers and make yourself very knowledgeable of what is available. You can profit nicely from the sale of cabinets, but you have to know what you are talking about to be fair to your customers and to turn the most profit for yourself.

CABINET OPTIONS

Cabinet options are nearly endless. I've recommended that you gather catalogs to educate yourself with. In the meantime, I am going to provide you with a long line of illustrations to get you started. Don't depend only on the selections I have included here. Look at various manufacturers and see what is currently available to you and your customer (Figures 4.12 through 4.43).

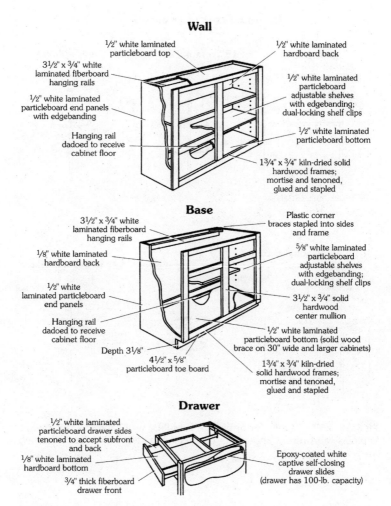

FIGURE 4.11 Detailed cross section of what to look for in quality cabinets. *(Courtesy of Wellborn Cabinet, Inc.)*

Diagonal corner peninsula

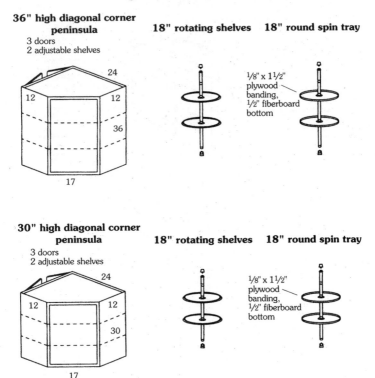

36" high diagonal corner peninsula
3 doors
2 adjustable shelves

18" rotating shelves   **18" round spin tray**

1/8" x 1 1/2" plywood banding, 1/2" fiberboard bottom

30" high diagonal corner peninsula
3 doors
2 adjustable shelves

18" rotating shelves   **18" round spin tray**

1/8" x 1 1/2" plywood banding, 1/2" fiberboard bottom

FIGURE 4.12 Details for appliance cabinets, wall cabinets, microwave cabinets, and return-angle wall cabinets. *(Courtesy of Wellborn Cabinet, Inc.)*

 Trade Tip: Don't rip out a kitchen until you have replacement cabinets in hand. Either store them in your facility, have your supplier hold them for you, get your cabinetmaker to keep them for you, or store them on site, but don't assume they will be available when you need them unless you can see them.

Diagonal corner appliance cabinets

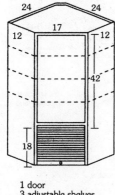
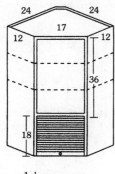

FIGURE 4.13 Diagonal corner appliance cabinets. *(Courtesy of Wellborn Cabinet, Inc.)*

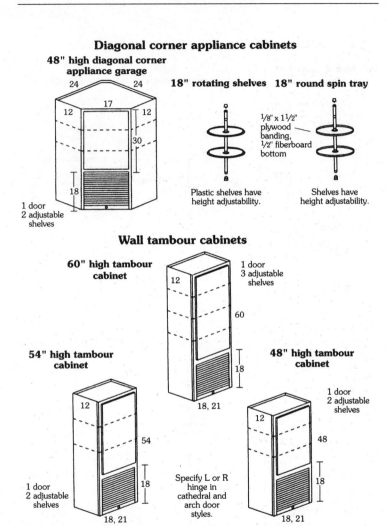

FIGURE 4.14 Diagonal corner peninsula cabinets that give customers superior storage in a small space. *(Courtesy of Wellborn Cabinet, Inc.)*

42" high microwave cabinet
2 doors
1 adjustable shelf
1 17¼" deep removable shelf.

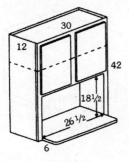

36" high microwave cabinet
2 doors
1 17¼" deep removable shelf.

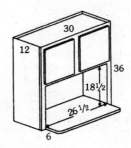

48" high microwave cabinet
2 doors
1 adjustable shelf
1 17¼" deep removable shelf.

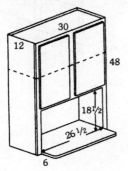

Base/wall return angle
2 doors
2 fixed shelves

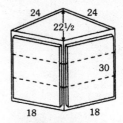

Inside angle 140°.
Outside angle 220°.
Square doors standard.
Ends not finished.

FIGURE 4.15 Diagonal corner peninsula cabinets that give customers superior storage in a small space. *(Courtesy of Wellborn Cabinet, Inc.)*

Corner base cabinets

36" corner base cabinet

42" corner base cabinet

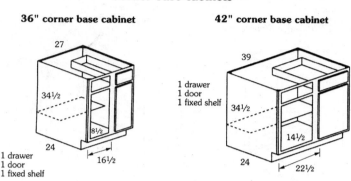

39" corner base cabinet

45" corner base cabinet

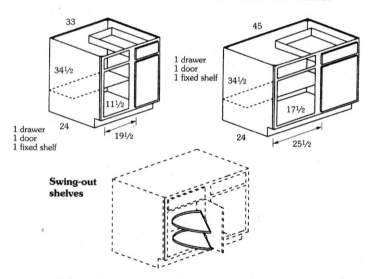

FIGURE 4.16 Options for corner base cabinets. *(Courtesy of Wellborn Cabinet, Inc.)*

Base accessories

Silverware tray

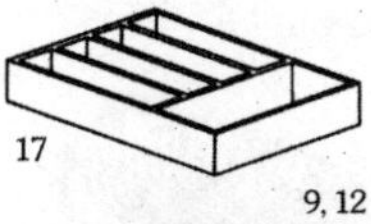

Silverware divider

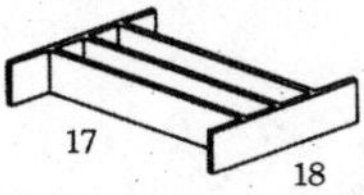

Drawer spice rack

White high-density polyethylene. May be trimmed.

Silverware tray

Double-tiered white tray. Guides have epoxy-coated sides. Top drawer integrates with bottom drawer by sliding back to reveal bottom drawer. Side flanges may be trimmed. Back of drawer must be removed for tray to be functional.

Interior drawer		
Min. height	Width	Min. depth
3½"	12¼ to 14¾"	17¼"

Utensil drawer kit
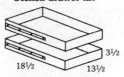

Slides are ⅜" oak. Bottom is 3/16" plywood. Kit includes 2 shelves and guides for installation.

Sliding shelf kit
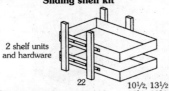

2 shelf units and hardware

⅝" oak sides, front and back
3/16" hardwood bottom.
Adjustable wooden mounting brackets.

Single roll-out shelf
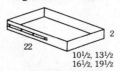

One shelf and guides for installation.
½" laminated particleboard sides.
¼" oak front. Bottom is ⅛" hardboard.

Cutting board

Includes cutting board, knife divider and drawer. Cutting board will replace existing drawer.

Double-tiered cutlery/cutting board

Includes cutting board and cutlery divider. Furnished with two hinges and necessary screws for installation. White cutting board is made of non-skid, non-absorbent polystyrene and is removable. Dishwasher safe. Side flanges may be trimmed.

Interior drawer		
Min. height	Width	Min. depth
3½"	18¼ to 20¼"	17¼"
3½"	12¼ to 14¼"	17¼"

FIGURE 4.17 Options for corner base cabinets. *(Courtesy of Wellborn Cabinet, Inc.)*

Base accessories

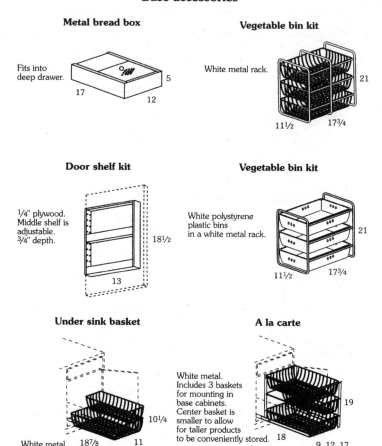

FIGURE 4.18 Options for corner base cabinets. *(Courtesy of Wellborn Cabinet, Inc.)*

Base acessories

Pull-out towel box

3 slide-out prongs for towels.
Mounts in all base cabinets.
White metal.

Ironing board

Built-in ironing board fits into drawer cavity and conveniently pulls out and unfolds.

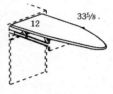

Single wastebasket

White wastebasket has 30 quarts of storage.
Guides have epoxy-coated sides. Full extension.
Includes mounts and screws.

Double wastebasket

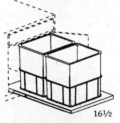

White wastebasket have 60 quarts of storage.
Guides have epoxy-coated sides. Full extension.
Includes mounts and screws.

3-bin recycle center

3-25 quart plastic bins. 1-18 quart cloth bin.
Equipped to separate recyclable materials: plastic, glass, paper, and metals.
Full extension. Includes mounts and screws.

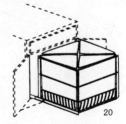

FIGURE 4.19 Options for corner base cabinets. *(Courtesy of Wellborn Cabinet, Inc.)*

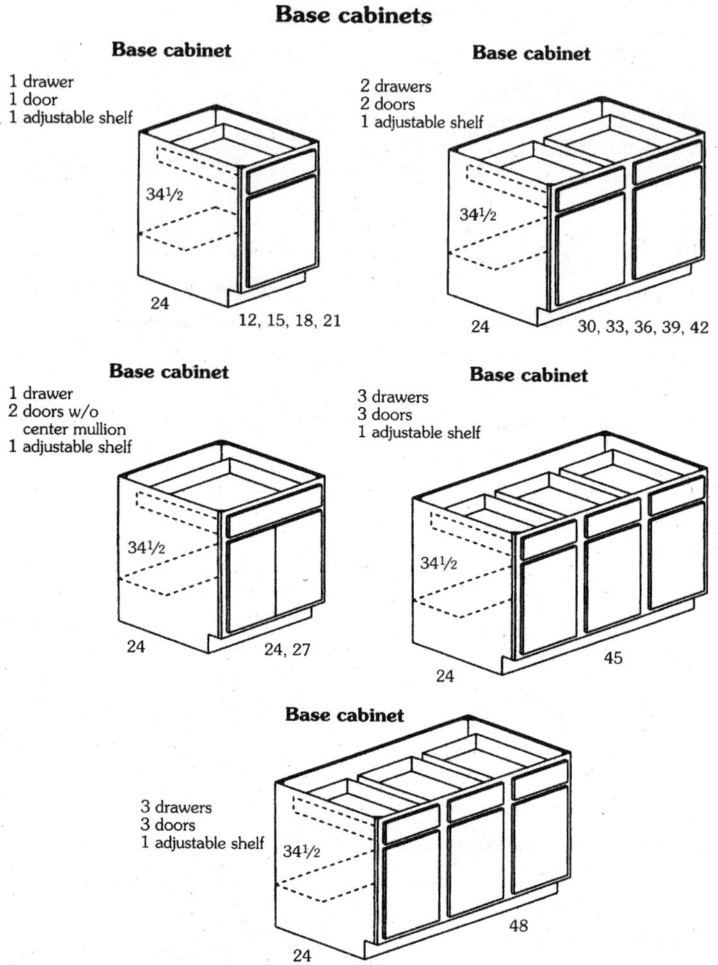

Base cabinets

Base cabinet
1 drawer
1 door
1 adjustable shelf

34½
24
12, 15, 18, 21

Base cabinet
2 drawers
2 doors
1 adjustable shelf

34½
24
30, 33, 36, 39, 42

Base cabinet
1 drawer
2 doors w/o center mullion
1 adjustable shelf

34½
24
24, 27

Base cabinet
3 drawers
3 doors
1 adjustable shelf

34½
24
45

Base cabinet
3 drawers
3 doors
1 adjustable shelf

34½
24
48

FIGURE 4.20 Options for corner base cabinets. *(Courtesy of Wellborn Cabinet, Inc.)*

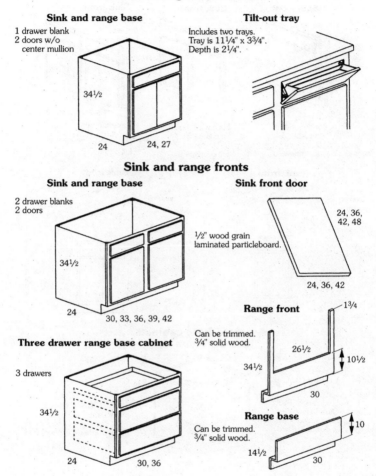

FIGURE 4.21 Options for corner base cabinets. *(Courtesy of Wellborn Cabinet, Inc.)*

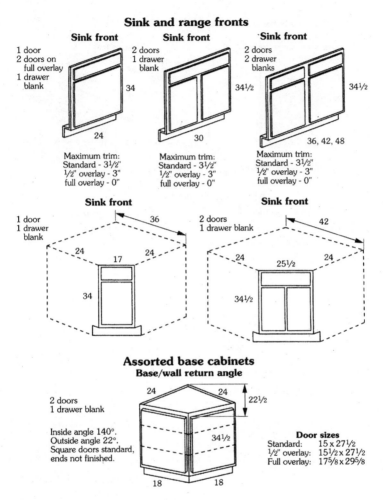

FIGURE 4.22 Sink and range fronts. *(Courtesy of Wellborn Cabinet, Inc.)*

Wall cabinets

30" high
1 door
2 adjustable shelves

9, 12, 15, 18, 21

24" high
2 doors w/o center mullion
1 fixed shelf

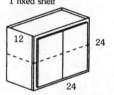

30" high
2 doors w/o center mullion
2 adjustable shelves

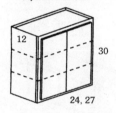

24, 27

30" high
2 doors
2 fixed shelves

30, 33, 36, 39, 42

30" high
3 doors
2 fixed shelves

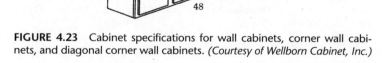

FIGURE 4.23 Cabinet specifications for wall cabinets, corner wall cabinets, and diagonal corner wall cabinets. *(Courtesy of Wellborn Cabinet, Inc.)*

Wall cabinets

24" high
2 doors
1 fixed shelf

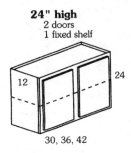

30, 36, 42

18" high
2 doors

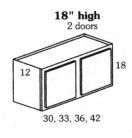

30, 33, 36, 42

18" high
1 door

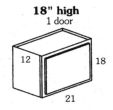

15" high
2 doors

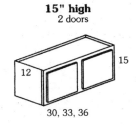

30, 33, 36

18" high
2 doors w/o center mullion

12" high
2 doors

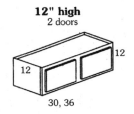

30, 36

FIGURE 4.24 Cabinet specifications for wall cabinets. *(Courtesy of Wellborn Cabinet, Inc.)*

Corner wall cabinets

30" high corner
2 doors
2 fixed shelves

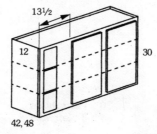

Corner wall cabinets

30" high corner
1 door
2 fixed shelves

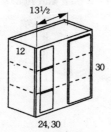

30" high corner
2 doors w/o center mullion
2 fixed shelves

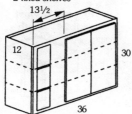

Diagonal corner wall cabinets

30" high diagonal corner
1 door
2 adjustable shelves

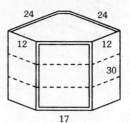

18" rotating shelves

FIGURE 4.25 Cabinet specifications for wall cabinets. *(Courtesy of Wellborn Cabinet, Inc.)*

Extended stiles

84" oven cabinet with 6" right stile

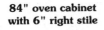

6" stile

84" utility cabinet with 3" left stile

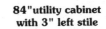

3" stile

30" wall cabinet with 3" left stile

27" base cabinet with 6" right stile

30" wall cabinet with 3" left and right stile

12" - 30" high walls, bases, vanities
36" high walls
42" high walls
84", 90" and 96" high utility, oven
3" per side
6" per side

Face cabinet in specifying left or right

Increased-depth wall cabinet

30" wall cabinet with 18" depth

18" depth

Up to 24" deep.

30" wall cabinet with 24" depth

24" depth

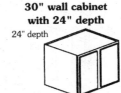

FIGURE 4.26 Examples of customized stock cabinet specifications that allow for extended stiles, increased depth, and other options. *(Courtesy of Wellborn Cabinet, Inc.)*

Cabinet depth reduction

84" utility cabinet with 8" depth

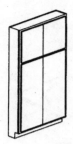

Wall cabinet: minimum depth of 8"
Base cabinet: minimum depth of 12"
Please note: by reducing cabinet depth, the following drawer features will be eliminated:
1. Self closing feature
2. Tightness of front frame
3. Drawers will need to be realigned when installed

Utility cabinet: minimum depth of 8"

Base cabinet with 16" depth

Wall cabinet with 8" depth

Peninsula cabinets

Peninsula wall cabinet

Peninsula base cabinet

Refrigerator cabinets

Full-height doors

Base cabinet with full-height doors

Base cabinet with full-height doors

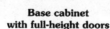

Mullion doors

FIGURE 4.27 Examples of customized stock cabinet specifications that allow for extended stiles, increased depth, and other options. *(Courtesy of Wellborn Cabinet, Inc.)*

Corner sink base cabinets

Adjustable shelf
2 shelves

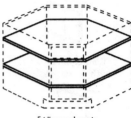

5/8" wood grain laminated particleboard.

28" Round spin tray

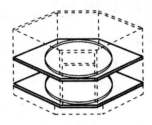

Wooden rotary shelf kit is 1/2" fiberboard with 1/8" x 1 1/2" plywood edgebanding.

Rotary kidney

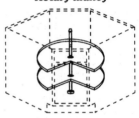

Shelves have height adjustability. 28" diameter. Shelves are made of plastic.

Base lazy susan
Double doors attached to rotary shelves

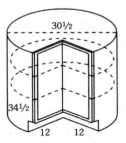

28" diameter.

28" Rotary shelves

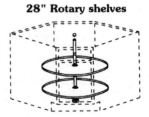

Shelves have height adjustability. 28" diameter. Shelves are made of plastic.

FIGURE 4.28 Corner sink base cabinets. *(Courtesy of Wellborn Cabinet, Inc.)*

Corner base cabinets

Corner base cabinet installation instructions (reversible)

Drawer track comes installed on door side only. To reverse corner cabinet move drawer track to other side.

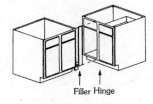

Filler Hinge

Left blind corner base cabinet is shown. Hinge is on blank side of cabinet.

	Standard	1/2" overlay	Full overlay
maximum pull	40¾	40¾	40¾
minimum pull w/hardware	37	37¼	40
minimum pull w/o hardware	36	36¼	37
maximum pull	43¾	43¾	43¾
minimum pull w/hardware	40	40¼	43
minimum pull w/o hardware	39	39¼	40
maximum pull	46¾	46¾	46¾
minimum pull w/hardware	43	43¼	46
minimum pull w/o hardware	42	42¼	43
maximum pull	49¾	49¾	49¾
minimum pull w/hardware	46	46¼	49
minimum pull w/o hardware	45	45¼	46

On all full overlay door styles use at least 1" base filler.

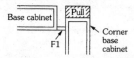

36" Corner base cabinet
1 drawer
1 door
1 adjustable shelf

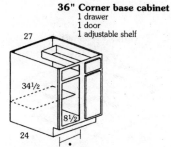

- Standard door = 16½
  1/2 overlay = 16¼
  Full overlay = 15³⁄₁₆

39" Corner base cabinet
1 drawer
1 door
1 adjustable shelf

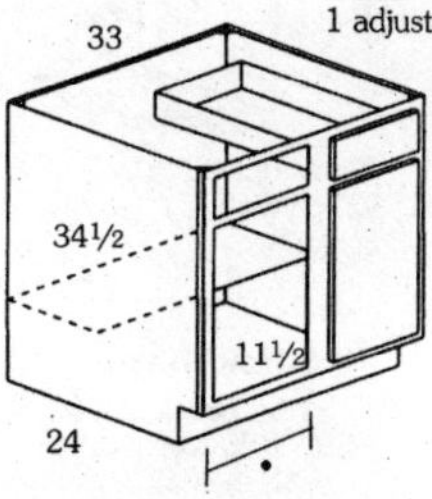

- Standard door = 19½
  1/2 overlay = 19¼
  Full overlay = 18³⁄₁₆

FIGURE 4.29 Corner base cabinets. *(Courtesy of Wellborn Cabinet, Inc.)*

Assorted base cabinets

Desk file drawer cabinet
1 drawer
1 file drawer

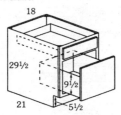

This file drawer will accommodate a Pendaflex file system. Specify Square or Cathedral door when ordering. File drawer is 1/2" solid wood. File drawer is 18 1/2" deep and 13 1/2" wide.

Corner sink base
Double doors
on piano hinge

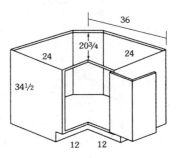

NOTE: Install shelves before countertop installation.

Wastebasket cabinet
1 full height door

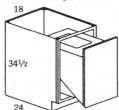

11 gallon capacity. Slides out 22 1/4".
Trash bin holder is 1/2" plywood.

Adjustable shelf
2 shelves

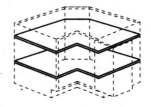

5/8" wood grain laminated particleboard.

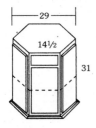

Island base
6 doors
1 drawer
1 shelf

Base molding
1 1/2" high.
Cut countertop
moldings at 30°.

FIGURE 4.30 Assorted base cabinets. *(Courtesy of Wellborn Cabinet, Inc.)*

Corner sink base cabinets

Rotary kidney

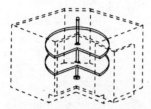

Shelves have height adjustability. 28" diameter. Shelves are made of plastic.

Corner recycle center

3–32 quart white plastic bins.

Super lazy susan kit

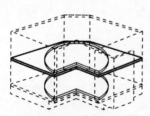

Wooden rotary shelf kit is ½" fiberboard with ⅛" x 1½" plywood edgebanding.

Diagonal corner sink base
1 door
1 drawer blank

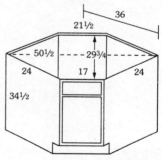

NOTE: Install shelves before countertop installation.

FIGURE 4.31 Corner sink base cabinets. *(Courtesy of Wellborn Cabinet, Inc.)*

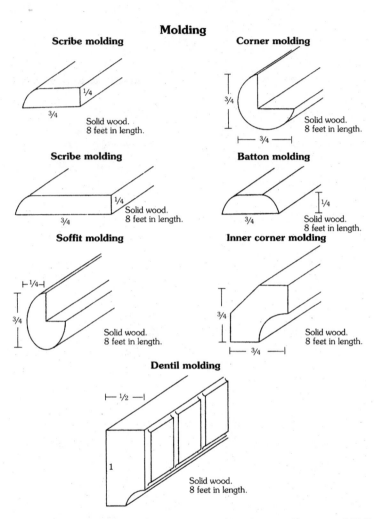

FIGURE 4.32 Types of moldings used with cabinets. *(Courtesy of Wellborn Cabinet, Inc.)*

Utility storage cabinets

Pantry storage kit

24 x 84 utility storage cabinet

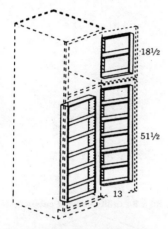

Allow 6" on both sides of pantry cabinet for door opening. 3/4" plywood kit with 1/2" plywood adjustable shelves.

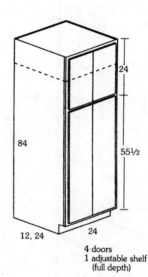

4 doors
1 adjustable shelf
(full depth)

Top view

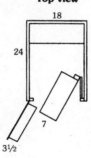

FIGURE 4.33 Utility storage cabinets. *(Courtesy of Wellborn Cabinet, Inc.)*

Utility storage cabinets

24 x 90 utility storage cabinet

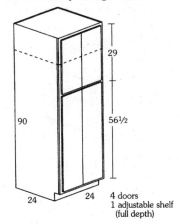

Utility sliding-shelf kit

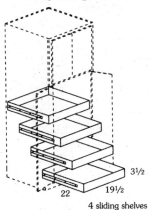

4 sliding shelves

24 x 96 utility storage cabinet

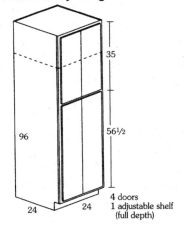

36 x 84 utility storage cabinet

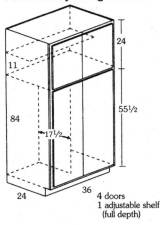

FIGURE 4.34 Utility storage cabinets. *(Courtesy of Wellborn Cabinet, Inc.)*

Utility storage cabinets

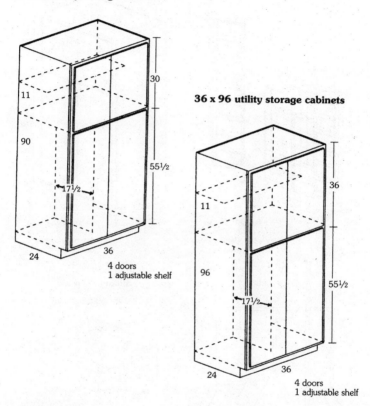

FIGURE 4.35 Utility storage cabinets. *(Courtesy of Wellborn Cabinet, Inc.)*

Utility storage cabinets

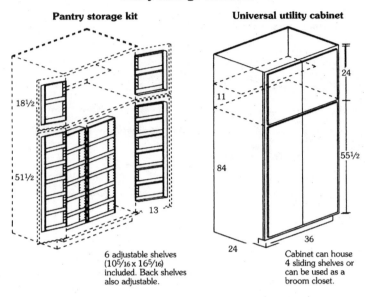

Pantry storage kit

6 adjustable shelves (10⁵⁄₁₆ x 16⁵⁄₁₆) included. Back shelves also adjustable.

Universal utility cabinet

Cabinet can house 4 sliding shelves or can be used as a broom closet.

Top view

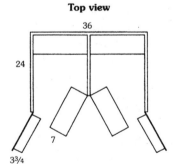

Sliding shelf kit

2 shelves and hardware

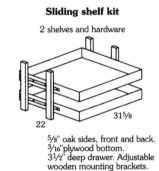

⅝" oak sides, front and back.
³⁄₁₆" plywood bottom.
3½" deep drawer. Adjustable wooden mounting brackets.

FIGURE 4.36 Utility storage cabinets. *(Courtesy of Wellborn Cabinet, Inc.)*

Base accessories

Cutting board

Includes cutting board, knife divider and drawer. Cutting board will replace existing drawer. Use existing drawer front with two hinges that are supplied. Nesessary harware is included also. Cutting board kit is made of oak wood.

Double tiered cutlery/cutting board

Includes cutting board and cutlery divider. Furnished with two hinges and necessary screws for installation. White cutting board is made of nonskid, nonabsorbant polystyrene and is removable. Dishwasher safe. Side flanges may be trimmed.

Interior drawer		
Min. height	Width	Min. depth
3½"	18¼" to 20¾"	17¼"
3½"	12¼" to 14¾"	17¼"

Drawer spice rack

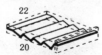

White high-density polyethylene. May be trimmed.

Silverware tray

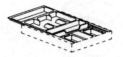

Double-tiered white tray.
Guides have expoxy-coated sides.

Top drawer integrates with bottom drawer by sliding back to reveal bottom drawer. Side flanges may be trimmed.
Back of drawer must be removed for tray to be functional.

Interior drawer		
Min. height	Width	Min. depth
3½"	12¼" to 14¾"	17¼"

Ironing board

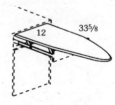

Built-in ironing board fits into drawer cavity and conveniently pulls out and unfolds.

FIGURE 4.37 Base cabinet accessories. *(Courtesy of Wellborn Cabinet, Inc.)*

Base accessories

Vegetable bin kit

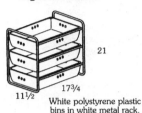

White polystyrene plastic bins in white metal rack.

Vegetable bin kit

White metal rack.

Double wastebasket

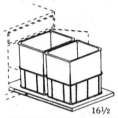

White wastebaskets have 60 quarts of storage.
Guides have epoxy-coated sides.
Full extension. Includes mounts and screws.

3-bin recycle center

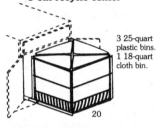

3 25-quart plastic bins.
1 18-quart cloth bin.

Equipped to separate the four most common recyclable materials: plastic, glass, paper and metals. Full extension. Includes mounts and screws.

Single wastebasket

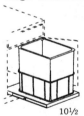

White wastebasket has 30 quarts of storage.
Guides have epoxy-coated sides.
Full extension. Includes mounts and screws.

Under-sink basket

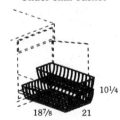

White metal. Top basket is removable.

FIGURE 4.38 Base cabinet accessories. *(Courtesy of Wellborn Cabinet, Inc.)*

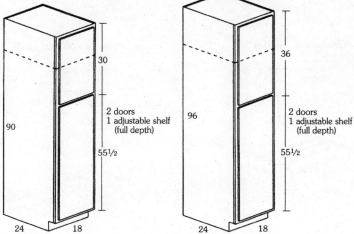

FIGURE 4.39 Utility storage cabinets. *(Courtesy of Wellborn Cabinet, Inc.)*

Base accessories

Silverware tray
1/4" oak.

Sliding shelf kit
5/8" oak sides, front and back.
3/16" hardwood bottom.
Adjustable wooden mounting brackets.

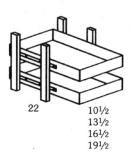

17   9,12

22   10 1/2
     13 1/2
     16 1/2
     19 1/2

Silverware divider
1/4" oak.
Can be trimmed to 15".

Single roll-out shelf
One shelf and guides for installation.
1/2" wood grain laminated particleboard sides.
5/8" oak front.
Bottom is 1/8" hardboard.

17   18

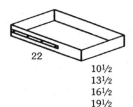

22   10 1/2
     13 1/2
     16 1/2
     19 1/2

FIGURE 4.40 Accessories for base cabinets. *(Courtesy of Wellborn Cabinet, Inc.)*

Base accessories

Utensil drawer kit
Sides are 3/8" oak.
Bottom is 3/16" plywood.
Kit includes 2 shelves and guides for installation.

Metal bread box
Fits into deep drawer.

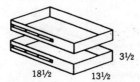

Door shelf kit
1/2" plywood.
Middle shelf is adjustable.
3 1/2" depth.

Bread box cover
Fits in drawer base 12"–24".
Can be trimmed.
Overall size is 17 1/8" x 21 1/4".
Mounts to drawer sides.

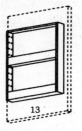
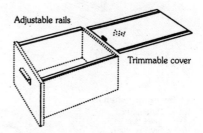

FIGURE 4.41 Accessories for base cabinets. *(Courtesy of Wellborn Cabinet, Inc.)*

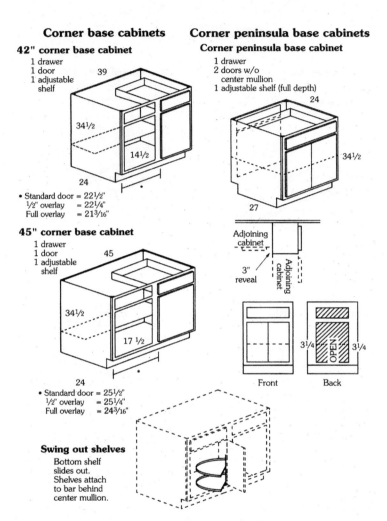

FIGURE 4.42 Corner base cabinets. *(Courtesy of Wellborn Cabinet, Inc.)*

Base whatnots

Peninsula base whatnot
12" depth
4½" matching toe kick
¾" solid wood shelves

34½

Detachable toe kick
for wall installation.

Base end shelf
12" radius.
4½" matching toe kick.
¾" solid wood shelves.

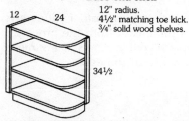

Detachable toe kick for wall installation
and L or R reversibility.

L-Shaped corner base shelf
24" depth
4½" matching toe kick
¾" solid wood shelves

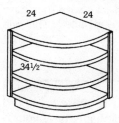

Detachable toe kick for wall installation.

Peninsula base cabinets

Peninsula base cabinet
1 drawer
1 drawer blanks
2 doors on both sides
  w/o center mullion
1 adjustable
  shelf (full depth)

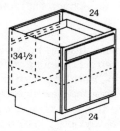

Peninsula base cabinet
2 drawers
2 drawer blanks
2 doors on both sides
1 adjustable
  shelf (full depth)

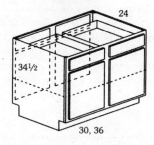

FIGURE 4.43 Corner base cabinets. *(Courtesy of Wellborn Cabinet, Inc.)*

USING EXISTING CABINETS

Using existing cabinets is one way for homeowners to save money when remodeling their kitchens. However, few homeowners will want to keep their existing cabinets in their present condition. This usually means refinishing or refacing the cabinets. This is an economical option that is, in my opinion, best left to experts.

Your crews may well be qualified to reface or refinish cabinets. However, this can be a tricky business and sometimes results in costly errors. I would prefer to hire a subcontractor for this type of work, but the choice is yours.

Refurbishing existing cabinets is quite common. If your crews are up to the task, you can turn some cash from the process. I have always preferred to use replacement cabinets as insurance against callbacks and potentially angry customers.

INSTALLING CABINETS

Don't cut corners when installing cabinets. I have far too many bad memories of workers who were lazy or in a hurry and having to pay the price later. It's doesn't take long to install base cabinets so that they appear to be okay. But, when you go to install the countertop, you may well find that the cabinet installation is not nearly as good as you thought it was. Make sure that base cabinets are level and plumb when they are installed. This will save you a lot of grief later.

Trade Tip: If you are going to reuse existing cabinets in an extensive remodeling job, have your customer go through a careful inspection of the cabinets with you before any work is done. Note any existing defects. It's a good idea to photograph the existing cabinets. Once you and your customer have completed the inspection, have the customer sign off on the notes you have made. This may save you from being blamed for existing defects.

Wall cabinets can also be deceptive when you first look at them. Insist on close supervision when having cabinets installed by someone else. As the remodeling contractor, the ultimate responsibility is yours. It is not only expensive to have to remove and reinstall new cabinets, it is also embarrassing. Your reputation is at stake.

COUNTERS

Counters are sometimes thought of as the crowning glory of a kitchen. The decisions required to pick the perfect counter are numerous. Laminated counters are the most common. Marble counters are sometimes used, but these are very heavy and very expensive. There are some simulated marble counters that work very nicely and customers tend to love them. Counters covered with ceramic tile are also extremely popular with some customers.

When planning a counter, you must consider how much backsplash the counter will have. It's typical to have a fairly low backsplash. Some owners like higher backsplashes. In fact, it is not uncommon to tile a section of wall above a counter to create a large backsplash.

When measuring for a counter, make sure that your measurements are correct. A minor mistake with the tape measure can result in a scrapped countertop. You must be precise when ordering and installing countertops.

PLANNING

Planning is a major factor in successful kitchen remodeling. All remodeling requires planning and good organizational skills. The level of expertise in these areas is elevated when working with kitchens. People don't like to be without their kitchen conveniences for long periods of time. Having cabinets or counters held up for delivery can ruin your plans. This can be overcome by scheduling your work properly and being certain that materials will be ready for you when you need them. If you plan your work and work your plan, you will find that kitchen remodeling can be very lucrative.

QUESTIONS AND ANSWERS

Q: What makes kitchen remodeling so popular?
A: *One thing that makes kitchen remodeling popular is that kitchens get a lot of use and people want them to be user-friendly. Another key reason for the popularity of kitchen remodeling is the rate of return a homeowner gets on the investment in a remodeling job.*

Q: How important is it to only use custom cabinets?
A: *Custom cabinets are not very important in the overall picture. They have their place and are very necessary in high-end, custom jobs. But, high-quality stock cabinets can do very well for your reputation.*

Q: Are appliances usually considered a part of a remodeler's responsibility?
A: *There is no standard answer. Your answer will be found in your agreement with your customer. Certainly, appliances are a needed element of a kitchen and they should be addressed in the contract between you and your customer.*

Q: How important is overhead lighting in a kitchen?
A: *My experience has shown that customers like bright kitchens. I feel that overhead lighting is very important, but I think that the fixtures should be chosen carefully.*

Q: Is wallpaper really suitable for use in a kitchen?
A: *As long as the wallpaper can be scrubbed clean and is resistive to steam and moisture, it should be fine.*

Q: Is a permit needed to replace a kitchen sink and dishwasher?
A: *If the location of the fixtures is not being changed, a permit should not be needed.*

Q: Should I laminate my own counters?
A: *If you have the proper skills and tools, you can. However, I prefer to buy finished counters for my remodeling jobs.*

Q: How do you feel about sculptured counters that are not the typical rectangular style?
A: *I like them. Using a counter with sweeping designs can open up the work area of a kitchen and it can also add modern flair to the overall look of a kitchen.*

Chapter 5

BATHROOM REMODELING

Bathroom remodeling is big business. There are contractors who specialize in bath and kitchen remodeling and who don't take other types of jobs. Without question, there is a huge demand for qualified remodelers who work wonders with bathrooms. The work can be as simple as replacing a couple of fixtures and refinishing a bathtub, or the work can mean expanding the size of a bathroom and moving or adding fixtures. Then there are the times when a bathroom is installed where there wasn't one. It's quite common to tuck a half-bath under a stairway or to put a master bath in a walk-in closet. The amount of skill and work required for these different types of approaches can be significant.

We have already talked about basement bathrooms. This chapter is going to deal with the renovation of existing bathrooms and the installation of new bathrooms in existing space. The nature of the work between these two types of remodeling is considerable.

REPLACING EXISTING FIXTURES

If all you are doing is replacing existing fixtures, bathroom remodeling is fairly easy. However, the bathing unit presents some problems under these circumstances. Toilets and bidets can be replaced quickly and easily. When these fixtures are changed, you might have to replace the bathroom flooring. This can be the case when new fixtures have a smaller footprint than existing fixtures. Lavatories and vanities are not difficult to replace, but you may have some work to do on the walls to hide defects from the change of fixtures.

Bathtubs and showers are more complicated to replace. To remove these fixtures, you have to create some wall damage. Refinishing a bathtub is one way to make it look fresh without creating wall damage or increased remodeling costs. Let's look briefly at the subject of removal of bathing units (Figure 5.1).

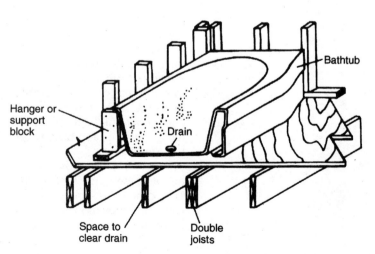

FIGURE 5.1 Typical rough installation of a bathtub.

Bathtubs often have ledges on them that are concealed by walls. In any event, a portion of the wall above the rim of a bathtub must be removed before the tub can be removed. The same type of ledge is present on most shower bases. Anyway you look at it, some wall damage is going to be incurred when removing a typical bathing unit. You may be able to repair or replace the wall section after a new bathing unit is installed, but matching the repaired or replaced section of wall with existing wall finishes can be nearly impossible. It is far safer to plan on having to put a new finish on all wall surfaces to ensure a quality job.

Also, don't attempt to replace a bathtub with a shower, unless you are prepared to install a new drain for the shower. Bathtubs require a minimum drain diameter of 1½ inches, even when a showerhead is installed above the bathtub. Showers require a drain with a minimum diameter of 2 inches (Figure 5.2).

When all you are required to do is to replace plumbing and light fixtures in a bathroom, you've got a fairly easy job. You may well wind up replacing the floor covering and wall finishes, but it's still not a bad job. And, you can make a lot of money in a relatively short period of time.

FULL BATHROOM REMODELING

Full bathroom remodeling can be a lot to bite off, but it is very manageable if you plan your work and work your plan. Complete bathroom remodeling can involve replacing rotted subflooring, repairing damaged floor joists, running new electrical wiring, and so forth. You have to assess a job carefully when you are bidding it.

Did You Know: that a plumbing permit is not required when plumbing fixtures are simply being replaced with new fixtures in the existing location? Well, you don't permit for this, but you will need a permit if you change fixture locations or add new fixtures to a plumbing system.

Type of fixture	Size (in.)
Clothes washer	2
Bathtub with or without shower	1½
Bidet	1½
Dental unit or cuspidor	1¼
Drinking fountain	1¼
Dishwasher (domestic)	1½
Dishwasher (commercial)	2
Floor drain	2, 3, or 4
Lavatory	1¼
Laundry tray	1½
Shower stall (domestic)	2
Sinks:	
Combination, sink and tray (with disposal unit)	1½
Combination, sink and tray (with one trap)	1½
Domestic with or without disposal unit	1½
Surgeon's	1½
Laboratory	1½
Flushrim or bedpan washer	3
Service	2 or 3
Pot or scullery	2
Soda fountain	1½
Commercial, flat rim, bar, or counter	1½
Wash circular or multiple	1½
Urinals:	
Pedestal	3
Wall-hung	1½ or 2
Trough (per 6-foot section)	1½
Stall	2
Water closet	3

FIGURE 5.2 Common trap sizes.

Some Items to Consider When Bidding Bathroom Remodeling

✔ Will the floor covering be replaced?
✔ Are there any signs of water damage or rot in the flooring system?
✔ Will the bathing unit be replaced?
✔ Will the bathing unit be refinished?

✔ Will a vanity be installed?
✔ What is the rough-in measurement on the existing closet flange?
✔ How does the existing bathroom layout in terms of fixture spacing?
✔ Does the bathroom have a ground-fault-interceptor circuit?
✔ Is there a window in the bathroom that opens to open air?
✔ Does ductwork exist for an exhaust fan?
✔ Will fixture locations be changed?
✔ What type of wall surfaces are available to work with?
✔ Will existing walls be torn out?
✔ What condition is the ceiling in?
✔ Will additional electrical work be needed?
✔ Does the bathroom need any heating work?
✔ Is there mold or mildew to be dealt with on walls or ceilings?
✔ What is the available access for bringing in a new bathing unit?
✔ Is there finished living space below the bathroom?

PLUMBING FIXTURES

Plumbing fixtures found in most bathrooms include a toilet, a lavatory, and a bathing unit. There are many types of fixtures that fall into these categories. Some other fixtures that might be encountered are bidets and whirlpool tubs. Customers are usually in charge of choosing plumbing fixtures; you and your plumbing contractor may have to bring the customers back to reality. I can't tell you the number of times that I have had customer requests that simply could not be met due to space limitations.

Lavatories

There are three basic types of lavatories to consider using in a bathroom. The least expensive, and usually the least desirable, type is a simple, wall-hung lavatory. This type of fixture is functional, but it is not pretty. If you plan to have a wall-hung lavatory installed, you must have a sturdy piece of backing in the wall cavity, behind the wallcovering, to bolt the lavatory hanger to. If this backing is not already in place, you will have to open the wall to install it.

Most homeowners prefer vanities with molded lavatory tops on them. These units are much more attractive and offer the utility of some storage space. Another advantage to using a vanity is that it can hide marks on a wall where an existing lavatory was removed. Vanities can be fitted with a wood counter top that is covered with an attractive finish, but molded tops that have the lavatory bowl as an integral part of the top are much more popular (Figure 5.3).

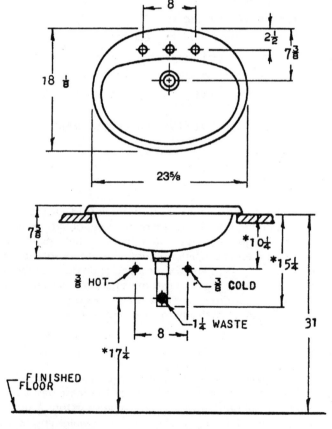

FIGURE 5.3 Self-rimming lavatory for use in a wooden counter.

Vanities are an element that some homeowners get carried away with. It's not unusual for a homeowner's eyes to be bigger than their bathroom is. Vanity cabinets come in a variety of sizes and styles (Figure 5.4). There are small ones for tiny bathrooms and very large ones that can accommodate two lavatory bowls, along with a host of drawers. There are even vanities made to accept the use of a chair for a sitting area to apply makeup and to take care of other personal business. The potential problem is space. No matter how badly a homeowner wants a 4 foot vanity, it just won't fit in a 2 foot space.

The third type of lavatory that is very popular is a pedestal lavatory (Figure 5.5). This type of lavatory is not nearly as practical as a vanity with a lavatory bowl in it, but the stylish look of a pedestal lavatory is pleasing to many customers.

Pedestal lavatories are a pain for plumber to install in a remodeling situation, but it can be done. What makes the installation difficult? The rough-in location of waste and water pipes for a pedestal lavatory have to be very close together. This is so that the pipes can be hidden by the pedestal. A typical rough-in will not work with a pedestal lavatory. So, a plumber will have to relocate existing piping in the wall to make it compatible with a pedestal lavatory.

Toilets

Toilets come in a wealth of colors and styles (Figure 5.6). When you get down to the basics, there are several considerations to take into account when choosing a toilet. Will your job have a conventional toilet or a toilet that is taller than a conventional toilet. The taller toilet is often referred to as a handicap toilet, but they are favored by many people who don't have physical disabilities. This is something to keep in mind.

Did You Know: that a pedestal lavatory requires the same type of in-wall backing that a wall-hung lavatory does? The backing is needed for a bracket that supports the lavatory. Once the pedestal is placed under the lavatory, the pedestal gives added support and hides the plumbing connections, but the bracket is needed.

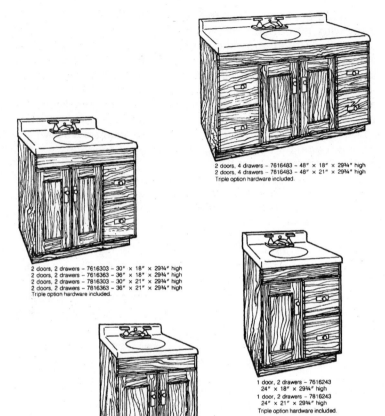

FIGURE 5.4 Types of vanities.

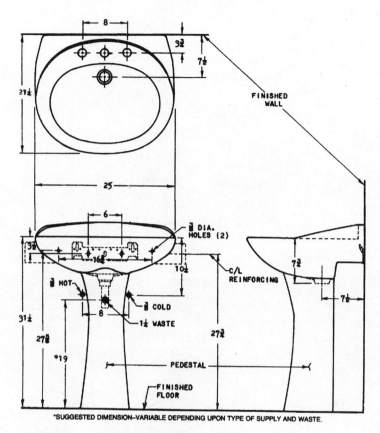

FIGURE 5.5 Pedestal lavatory.

  Trade Tip: If you decide to tighten the tank-to-bowl bolts or the flange bolts on a toilet, wear eye protection and don't stress the fixture. China will explode in your face if you stress it too much. At the least, tightening bolts too much will result in a crack that will leak.

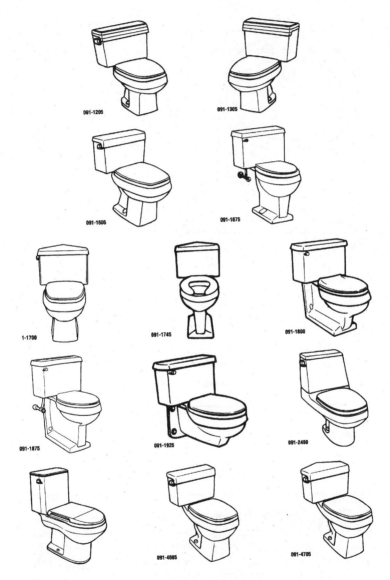

FIGURE 5.6 A variety of toilet types to choose from.

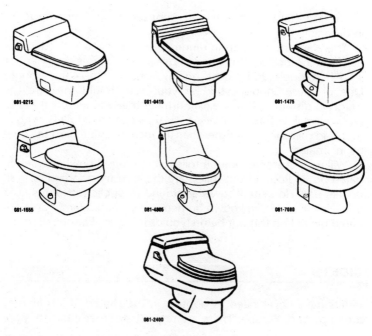

FIGURE 5.7 Low-profile, one-piece toilets.

The opposite of a high toilet is a low-profile toilet (Figure 5.7). These toilets are extremely popular when people are making a fashion statement with their bathroom fixtures. Functionally, the toilets are the same, but they have a different look. Many of these units are offered as one-piece construction. The tank and the bowl are all formed together as a single unit.

Between tall toilets and low toilets are standard toilets. These are the fixtures that are used most of the time. Even though these are called standard toilets, you can get them with a twist. Have you ever seen a corner toilet? They make them. The toilet tank is shaped to fit tightly into the corner of a room. I had one these in a house I built for myself years ago. If nothing else, they make an interesting conversation piece.

Another major consideration is the rough-in rating for the toilet you need. Standard rough-ins are set for the center of the closet flange to be 12 inches from the finished wall behind the toilet (Figure 5.8). But, floor joists don't always cooperate, and you might find that the closet flange is not set at the standard rough-in distance.

There are three rough-in ranges available for toilets. The standard is a 12 inch rough. You can get a toilet that is designed for a 10 inch rough, and there is one available with a 14 inch rough. One of these should work for you. You should measure the rough-in distance on any existing toilet that you are planning to replace well before you need its replacement. Toilets with a rough-in measurement other than 12 inches can be hard to come by in small towns.

Will your customer want a toilet with a round bowl or an elongated bowl? Round-front toilets are the industry standard for residential use. Elongated bowls are found in public restrooms, as a rule. However, I have known homeowners who preferred an elongated bowl. Iron this out before you order your plumbing fixtures.

BIDETS

Bidets have never been as popular in the United States as they are in parts of Europe. There are requests for them, but they are

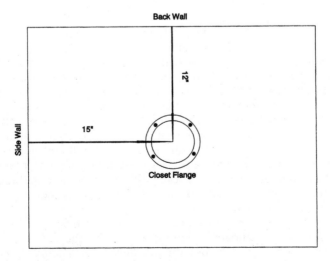

FIGURE 5.8 Standard toilet rough-in.

usually found only in high-end homes. I have had certain areas where I have worked and installed a number of bidets, but they have not been a staple fixture with my business. It is rare to find an old home with a bidet in it. Unless you are working an upscale market, I seriously doubt that you will have much call for bidets.

BATHING UNITS

Bathing units can be showers, bathtubs, or whirlpool tubs. Getting one-piece units into an existing home can be extremely difficult, if not impractical (Figure 5.9). Never fear, manufacturers offer many sectional units to overcome this problem. Standard bathtubs fit into a space that is 5 feet in length. Watch out for customers who spec out a 6 foot tub when you only have 5 feet to work with. This happens, so don't get taken by surprise if a customer chooses a model number that won't fit the available space.

If you need a sectional bathing unit, and you will for most jobs, there are plenty to choose from (Figure 5.10). Manufacturers make them in a number of ways. Most of them are a four-piece arrangement. Some of them come in just two pieces. There is no shortage of design styles or quality. I will tell you this: I have installed hundreds of sectional tub surrounds and you cannot afford to use a cheap one. Believe me on this. Pay what you have to pay to get the highest quality you can.

Custom-made showers are not very common in modern construction, but they are an option that you can offer your customers. These are showers where a plumber forms a base within the framing that your carpenters build. The base has a raised drain in it. Concrete, or a similar filler, is poured into the watertight base to create the rough shower floor.

 Don't Do This! Some customers may ask you to install functional handgrips in a shower or above a bathtub. Ask about this early on. If a solid handbar is wanted, you will do well to install some strong backing in the wall cavity before you seal up the wall.

In the old days, the liners were formed out of lead. Today, a special membrane material is used. Then a tile installer tiles the floor and walls. Seats can be installed, the size of the shower can be customized to fit on-site requirements, and the end result is a fully tiled shower that will get a lot of attention. Naturally, this type of shower is more expensive than a pre-formed base or a fiberglass unit, but if you have a customer who wants something different, this could be a great idea.

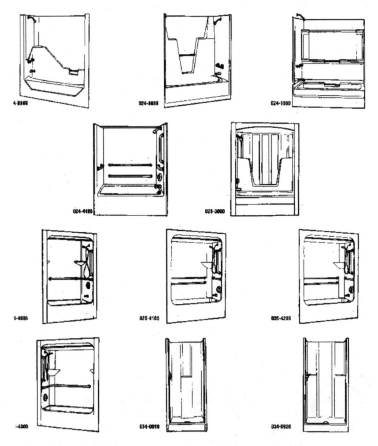

FIGURE 5.9 One-piece showers and tub-shower combinations.

#300 07-08-2016 3:18PM
Item(s) checked out to p10013490.

TITLE: Basic construction techniques for
BARCODE: 07367002438800
DUE DATE: 07-22-16

TITLE: Advanced framing methods : the il
BARCODE: 07367010036120
DUE DATE: 07-22-16

TITLE: Remodeler's instant answers
BARCODE: 07367010091687
DUE DATE: 07-22-16

TITLE: Ultimate guide to house framing :
BARCODE: 07367010561069
DUE DATE: 07-22-16

TITLE: Dream home : the Property Brother
BARCODE: 07367020034520
DUE DATE: 07-22-16

Driftwood Public Library
541-996-2277

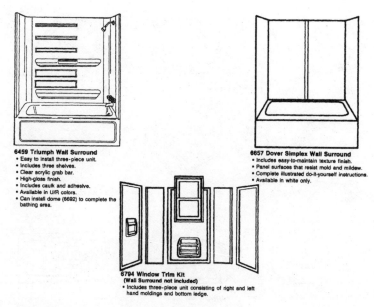

FIGURE 5.10 Sectional tub surrounds.

Whirlpool Tubs

Whirlpool tubs are very popular. There are models that will fit into the space of a standard bathtub. These models don't have a lot of jets or bells and whistles, but they will fit a standard space and get the job done (Figure 5.11). Most customers who want whirlpool tubs are thinking of larger units. Keep this in mind when estimating your jobs.

Whirlpool tubs come in a wide variety of sizes and styles. You can get whirlpool tubs that strongly resemble standard tubs, with the exception of the jets. Then there are units that are made to be installed so that the rim of the tub is nearly flush with the floor. For obvious reasons, these units are not suitable for second-floor installations. Many units are made to be placed in a framed enclosure that sits above floor level.

Size is one consideration when installing a whirlpool tub, but it is not the only factor to take into account.

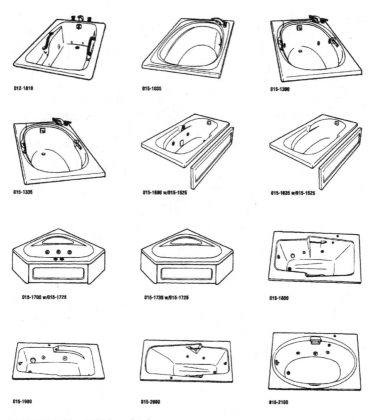

FIGURE 5.11 Whirlpool tubs.

WHIRLPOOL CONSIDERATIONS

✔ Will the tub fit in available space?
✔ Is the existing water heater serving the building large enough to provide suitable hot water when filling the whirlpool?
✔ Can the existing floor structure support the weight of the whirlpool when it is full of water and occupants?
✔ How will you create access to the motor and other components of the whirlpool unit that may require maintenance?

✔ Does the existing electrical service offer enough additional space for a circuit to provide electrical power to the whirlpool unit?
✔ What style of whirlpool does the customer want?

FAUCETS

Faucets can keep customers scratching their heads for days. Do they want single-handle faucets or two-handle faucets? Will pressure-balanced faucets be used with the bathing unit? Should the faucets have a chrome finish, an antique brass finish, or should they be plated with gold? A faucet selection for a lavatory can range in cost from under $50 to over $2,000. Yes, this is not a misprint. Gold faucets are available, some customers do buy them, and they are very expensive. Spec out your faucet list early and stick to it. Giving customers too much room to roam on this issue can be an expensive learning experience (Figure 5.12).

I find that the customers I have worked with in recent years tend to favor single-handle, chrome faucets. Some of them go for a more exotic finish, and a few go for the extremely expensive faucets. If you are installing a whirlpool tub, the faucet, or tub filler as it should be called, is going to be expensive. There is not a lot of advice need on this issue. You basically let your customers have what they want, just make sure you have it covered in your budget and that the faucet chosen will work on the fixture where it is intended to be installed.

Did You Know: that gold faucets can be damaged in the blink of an eye? They can. I had a plumber once who ruined an $1,800 lavatory faucet, and this was back in the early 80s. When I bought a second faucet of the same type, he managed to ruin that one too. It was an extremely expensive lesson for me to learn. Never trust super-expensive faucets to just any plumber.

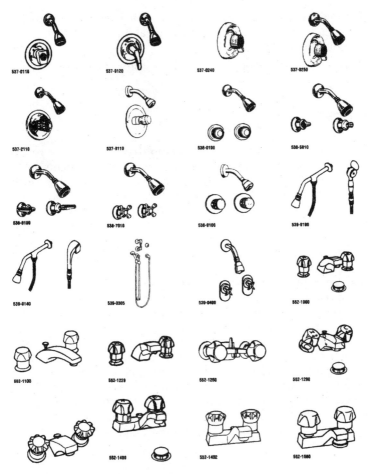

FIGURE 5.12 Samples of various faucet types.

MAKE A PLAN

Before you start to create a new bathroom or to expand the size of an existing bathroom, make a plan. Draw it up. If you can do it with a CAD program on your computer, that's great (Figure 5.13). When computers are not your forte, draw a plan on graph paper. One way or another, get a floorplan drawn for your customers to review and approve.

Let's say that you are going to take a large bedroom and frame walls inside of the bedroom to create a new bathroom (Figure 5.14). You need to show your customer what you intend to do before you take the job. If they like your plan, have them sign it and make it part of your remodeling contract. This is just good business. If a conflict arises during the job, you will have the signed plan to prove that you are doing what you were hired to do.

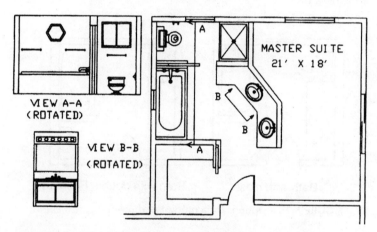

FIGURE 5.13 Typical CAD-drawn floorplan.

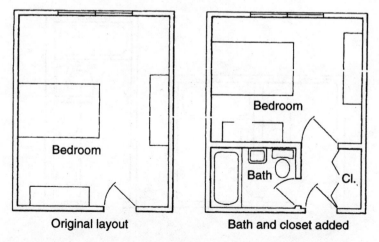

FIGURE 5.14 Adding a bathroom in a bedroom.

If your customers are not sure what they want, and many will not know what they want, give them some drawings to study and choose from (Figures 5.15 through 5.17). You don't have to create these drawings for every bid you give. Build an archive file of standard plans that you can show customers from job to job. Most of your competitors probably won't be doing this, so your actions will set you apart and may give you an advantage in winning the job.

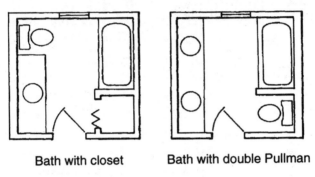

Bath with closet   **Bath with double Pullman**

FIGURE 5.15 Options for a modest bathroom.

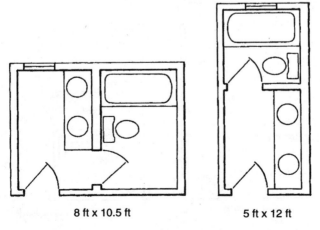

8 ft x 10.5 ft   **5 ft x 12 ft**

FIGURE 5.16 Drawing of a compartmentalized bathroom.

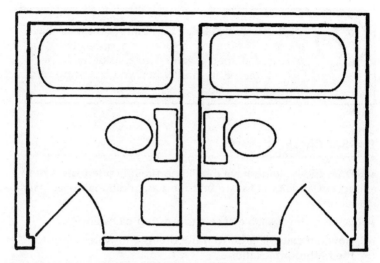

FIGURE 5.17 Back-to-back bathrooms that offer economical installation.

CONVERTING SPACE AND ADDING NEW BATHROOMS

Converting existing space and adding new bathrooms in the space can be a major job. This type of work can be far more advanced than standard bathroom remodeling. Code issues come into play. When you are dealing with buildings that have multiple stories, this can result in problems. Buildings that don't have access below the floor of the lowest level of living space complicate matters. This is not a job for rookies. But, it is good work with high profit potential.

Some homes and buildings have existing space that can be converted to a bathroom without a need for major framing and construction. Other situations require the bathroom space to be framed up and created. The framing is the easy part. Electrical work is usually not too difficult, but the plumbing elements of the bathroom can be challenging. Access and existing conditions have a lot to do with the level of difficulty involved in the remodeling process. You have to do your homework well before you bid these jobs.

Did You Know: that bathrooms must have either a window that opens to open air or an exhaust fan that is vented to outside air? This is fact, so keep it in mind when you are sizing up a bathroom addition.

Pre-Bid Checklist

✔ Does the building have a building drain of adequate size to accept additional fixture units and an additional toilet installation?
✔ What is the available access to existing plumbing?
✔ How difficult will it be to route drains and water pipes to the bathroom location?
✔ Can a 2 inch vent be installed without substantial damage to existing walls and ceilings?
✔ Will the existing electrical service accept the load of a new circuit?
✔ Does the proposed bathroom location have a window at this time?
✔ What will be required to install ductwork for an exhaust fan, if one is needed?
✔ Will the property owner accept electric heat in the new bathroom?
✔ Does the property owner want a heat lamp installed?
✔ What type of finished flooring will be installed?
✔ Will water-resistant drywall be used on the wall studs?
✔ What type of wall finish will be used?
✔ Is a painted ceiling wanted? Does the customer want a textured ceiling? Is some other type of ceiling preferred?
✔ What type and size of mirror, or mirrors, will be installed?
✔ Will a medicine cabinet be installed? If so, will it be a recessed cabinet or a flush cabinet.
✔ What type of bathing unit and faucet will be used? Will the mixing valve (faucet) be pressure balanced? What style and finish is required?

✔ Does the customer desire a vanity with a molded lavatory top? If not, what is wanted?
✔ What style and finish is wanted for the lavatory faucet?
✔ What style of toilet will be installed?
✔ How many light fixtures will be installed, and what type of fixture will they be?
✔ What color will the plumbing fixtures be?
✔ What style of door will be installed, and what will the door dimensions be? What type of hardware is wanted on the door?
✔ What type of finish trim will be installed? Will it be painted or stained?
✔ What accessories are wanted, such as towel racks, paper holders, shelves, and so forth?

The checklist above is only a sample. It is not conclusive, but it does give you a good idea of the types of questions to ask. Experience will give you the opportunity to develop your own, specialized checklists. You can almost never ask too many questions before you bid a job. While you may think that your questions will aggravate your customers, the questions are more likely to put them at ease. And, you will probably get some brownie points for being the only contractor asking them specific questions to ensure their satisfaction.

Trade Tip: Remodeling is a competitive business. You need an edge. Being very good at what you do is great, but you need to make your potential customers aware of your expertise. Some contractors try to do this with brochures and fancy advertisements. This can work, but word-of-mouth referrals are the best advertising that you can't buy. How do you get people to recommend your company? Give your customers more than they expect for less than they expect to pay.

Basic Carpentry

Building a new area for a bathroom addition doesn't require much more than basic carpentry skills. It is the plumbing, and sometimes the electrical work, that requires advanced skills. Framing stud walls, covering them with drywall, finishing the drywall, painting the walls, and installing a floor is all pretty standard stuff. Making the plumbing work is the hardest part. And, the electrical work can be tricky when access is limited, but electricians have a lot more viable options for routing their wires than plumbers have for installing their pipes.

Electrical Work

Electrical work for a new bathroom is usually pretty easy. As long as the existing service panel can handle the load of the new bathroom, a good electrician should be able to get wires to the location. The electrician can run wires under the building, if there is access, or run them through an attic to get them to the location. A possible glitch is if the new bathroom has finished living space below it and above it. This type of situation will test any electrician. But, an experienced electrician should be able to use closets and other similar routes to get wires to the new bathroom location. Once the wires are to the work area, wiring the walls is easy.

Electricians usually install the ductwork from exhaust fans to open air. This is much more difficult than working with a small wire. It's not a problem if there is an attic over the bathroom location. In some situations, a dropped ceiling can be used to hide the duct as it is routed to an exterior sidewall. This is a consideration that you must plan for in advance. If the bathroom location will have a window that opens to open air, the exhaust-fan duct is not needed.

Plumbing

Plumbing is the potential nightmare to adding a new bathroom. There are numerous code requirements to meet and plenty of potential obstacles to make the job seem less than worthwhile. Plus, good plumbers who have extensive remodeling experience don't come cheap.

Bathroom Remodeling

Trade Tip: I have built as many as 60 new homes a year, am a licensed master plumber, and have been involved with extensive remodeling projects for over 20 years. Having been in business for myself since 1979, I have learned a lot, most of it the hard way. Let me give you a piece of advice. When you are doing a complicated remodeling job, don't use tradespeople who do mostly new construction. There is no comparison between new work and remodeling. Sure, there are similarities and some trades can cross over more easily than others. I don't mean to point a finger at individual trades, but make sure that your subcontractors have what it takes to tackle complex remodeling work before you put them on the job.

Your plumbing contractor should go with you to estimate the price for putting a new bathroom in a building. This is not simple fixture exchange. Unless you know the plumbing code very well, you should not make assumptions on what will be needed, and plumbing is an expensive part of the job.

There are code requirements that dictate fixture spacing (Figure 5.18). This part of the code is pretty easy to understand. What you have to be concerned with are the more complicated issues and the practicality of getting plumbing to the new bathroom location.

Someone has to assess available plumbing connections. For example, if a home has a 3 inch sewer and already has two toilets, you are out of luck. A 3 inch sewer is not allowed to have more than two toilets connected to it. Installing a new toilet when a new sewer is required for it becomes incredibly expensive. You will also have to determine what is available to tap into for water pipes. As a rule of thumb, you cannot have more than two fixtures served by a ½ inch water pipe (Figure 5.19). This means that you should be looking for ¾ inch water piping to tie into.

Vents are required for all plumbing fixtures. The minimum size of the vents required depend on the type of fixture being vented. At a minimum, you have to have at least one 2 inch vent when a toilet is installed and a 3 inch vent is already existing somewhere in the plumbing system (Figure 5.20).

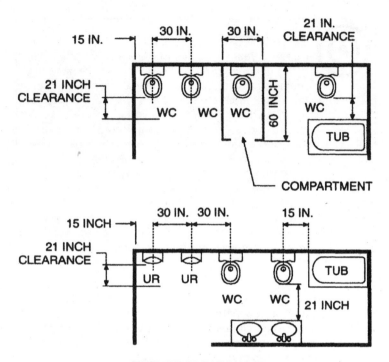

FIGURE 5.18 Recommended fixture clearances.

There are regulations that mandate how far a fixture trap may be located from its vent (Figure 5.21). This can be a factor when a recessed medicine cabinet it used. If the vent has to be kept out of the path of the medicine cabinet, the trap arm may extend beyond allowable distances.

I could give you chapter and verse of the plumbing code, but it is not your job to be a master plumber. It is, however, your job to be aware of what you might be getting into when you take on a remodeling job that requires the addition of a new bathroom. Please take my advice and have a good remodeling plumber with you when you estimate a job where a bathroom will be added. It's also a good idea to take an electrician with you. Property owners will appreciate the fact that you are bringing in expert

Fixture or device	Size (in.)
Bathtub	½
Combination sink and laundry tray	½
Drinking fountain	⅜
Dishwashing machine (domestic)	½
Kitchen sink (domestic)	½
Kitchen sink (commercial)	¾
Lavatory	⅜
Laundry tray (1, 2, or 3 compartments)	½
Shower (single head)	½
Sink (service, slop)	½
Sink (flushing rim)	¾
Urinal (1" flush valve)	1
Urinal (3/4" flush valve)	¾
Urinal (flush tank)	½
Water closet (flush tank)	⅜
Water closet (flush valve)	1
Hose bib	½
Wall hydrant or sill cock	½

FIGURE 5.19 Minimum size of fixture branch piping.

Type of fixture	Minimum size of vent
Lavatory	1¼"
Drinking fountain	1¼"
Domestic sink	1¼"
Shower stalls (domestic)	1¼"
Bathtub	1¼"
Laundry tray	1¼"
Service sink	1¼"
Water closet	2"

Note: At least one 3" vent must be installed.

FIGURE 5.20 Minimum vent sizes.

Size of fixture in inches	Distance trap to vent
1¼	2 ft. 6 in.
1½	3 ft. 6 in.
2	5 ft.
3	6 ft.
4	10 ft.

Note: Figures might vary with local plumbing codes.

FIGURE 5.21 Trap-to-vent distances.

before you even have the job. I know this to be true based on comments that past customers have made to me.

Drains

Drains are a major concern when adding a new bathroom. Assuming that your job site has an adequate building drain or sewer to connect to, you have part of the battle won. But, the building drain or sewer must be accessible and low enough to allow the new drains to be installed within code compliance (Table 5.1). Keep in mind that drains normally fall at a rate of ¼ inch for every foot that they travel. If there is a basement or crawlspace under the new bathroom location, a suitable drain connection, and enough grade to make the connection, you are well on your way to success. I can't turn you into a plumber in a few paragraphs, so take a pro on the estimate with you.

If the area below the proposed bathroom location is finished living space, you have a problem. Expect to tear out a portion of a wall to install suitable piping. Again, your plumbing contractor can tell you what is needed in this type of situation, but I can tell you that there is a good chance you will have to scrap a portion of a wall and repair it. This may mean repainting an entire room below the bathroom location.

The plumbing code dictates a lot of rules and regulations. It is unrealistic to think that a general contractor or remodeler is going to keep the plumbing code at the front of their mind during an estimate. You've probably done a lot of jobs. At what intervals do PVC drainage pipes have to be supported? How about PEX piping, do you know what the support intervals for this type of piping are? These are the simple questions. I'm trying to stress my point that you really owe it to yourself to have plumbers and electricians with you when you do on-site inspections for estimates (Table 5.2).

TABLE 5.1 Slope of Horizontal Drainage Pipe

SIZE (inches)	MINIMUM SLOPE (inch per foot)
2½ or less	¼
3 to 6	⅛
8 or larger	1/16

For SI: 1 inch = 25.4 mm, 1 inch per foot = 0.0833 mm/m.

Trade Tip: When you have to open a wall cavity, I suggest you do the initial opening with a hammer. Does this surprise you? It surprises a lot of people, but I have a good reason for the suggestion. How would you do it? Would you use a reciprocating saw? That's what a lot of remodelers would do. But, what happens if there are electrical wires in the wall cavity? Sparks will fly. If you break a channel in the drywall with a hammer, you can see what you are dealing with before you make the full and final cut in the wall section. It's safer and smarter.

Vents

Vents that are required for a new bathroom are not a big deal if there is attic space directly above the ceiling of the new bathroom. When living space is above the proposed bathroom location, the game gets a bit more interesting. Your plumber is going to have to get a 2 inch vent to either open-air termination or to a suitable connection point above the highest existing plumbing fixtures. This can be a considerable problem.

If you have finished living space above the proposed bathroom location, be prepared to open a wall cavity above the bathroom to allow for the venting of the bathroom. There is a chance that a seasoned remodeling plumber can get the pipe through a finished wall without having to tear the wall open, but you can't count on this when you are bidding a job. Plan for the worst and hope for the best.

Water Distribution Pipes

Water distribution pipes are needed when a new bathroom is added. It is best to find existing pipes that have a minimum

 Did You Know: that every building that contains a toilet is required to have at least one 3 inch vent that terminates in open air? Once this requirement is met, additionally toilets require venting with a pipe that has a minimal diameter of 2 inches.

TABLE 5.2 Hanger Spacing

PIPING MATERIAL	MAXIMUM HORIZONTAL SPACING (feet)	MAXIMUM VERTICAL SPACING (feet)
ABS pipe	4	10[b]
Aluminum tubing	10	15
Brass pipe	10	10
Cast-iron pipe[a]	5	15
Copper or copper-alloy pipe	12	10
Copper or copper-alloy tubing, $1\frac{1}{4}$-inch diameter and smaller	6	10
Copper or copper-alloy tubing, $1\frac{1}{2}$-inch diameter and larger	10	10
Cross-linked polyethylene (PEX) pipe	2.67 (32 inches)	10[b]
Cross-linked polyethylene/ aluminum/crosslinked polyethylene (PEX-AL-PEX) pipe	$2\frac{2}{3}$ (32 inches)	4
CPVC pipe or tubing, 1 inch or smaller	3	10[b]
CPVC pipe or tubing, $1\frac{1}{4}$ inches or larger	4	10[b]
Steel pipe	12	15
Lead pipe	Continuous	4
PB pipe or tubing	2.67 (32 inches)	4
Polyethylene/aluminum/polyethylene (PE-AL-PE) pipe	2.67 (32 inches)	4
PVC pipe	4	10[b]
Stainless steel drainage systems	10	10[b]

For SI: 1 inch = 25.4 mm, 1 foot = 304.8 mm.

a. The maximum horizontal spacing of cast-iron pipe hangers shall be increased to 10 feet where 10-foot lengths of pipe are installed.

b. Midstory guide for sizes 2 inches and smaller.

diameter of ¾ inch. Just as there are code regulations for drains and vents, there are minimum requirements for water distribution piping (Tables 5.3 through 5.5).

TABLE 5.3 Minimum Sizes of Fixture Water Supply Pipes

FIXTURE	MINIMUM PIPE SIZE (Inch)
Bathtubs (60" × 32" and smaller)[a]	1/2
Bathtubs (larger than 60" × 32")	1/2
Bidet	3/8
Combination sink and tray	1/2
Dishwasher, domestic[a]	1/2
Drinking fountain	3/8
Hose bibbs	1/2
Kitchen sink[a]	1/2
Laundry, 1, 2 or 3 compartments[a]	1/2
Lavatory	3/8
Shower, single head[a]	1/2
Sinks, flushing rim	3/4
Sinks, service	1/2
Urinal, flush tank	1/2
Urinal, flush valve	3/4
Wall hydrant	1/2
Water closet, flush tank	3/8
Water closet, flush valve	1
Water closet, flushometer tank	3/8
Water closet, one piece[a]	1/2

For SI: 1 inch = 25.4 mm, 1 foot = 304.8 mm, 1 psi = 6.895 kPa.

[a]. Where the developed length of the distribution line is 60 feet or less, and the available pressure at the meter is a minimum of 35 psi, the minimum size of an individual distribution line supplied from a manifold and installed as part of a parallel water distribution system shall be one nominal tube size smaller than the sizes indicated.

TABLE 5.4 Maximum Flow Rates and Consumption for Plumbing Fixtures and Fixture Fittings

PLUMBING FIXTURE OR FIXTURE FITTING	MAXIMUM FLOW RATE OR QUANTITY[b]
Water closet	1.6 gallons per flushing cycle
Urinal	1.0 gallon per flushing cycle
Shower head[a]	2.5 gpm at 60 psi
Lavatory, private	2.2 gpm at 60 psi
Lavatory (other than metering), public	0.5 gpm at 60 psi
Lavatory, public (metering)	0.25 gallon per metering cycle
Sink faucet	2.2 gpm at 60 psi

For SI: 1 gallon = 3.785 L, 1 gallon per minute = 3.785 L/m, 1 psi = 6.895 kPa.

a. A hand-held shower spray is a shower head.
b. Consumption tolerances shall be determined from referenced standards.

Water pipes are small enough that they are fairly easy to route to a new location. If they can be run in a crawlspace, a basement, or an attic, the task is simple. Should you have to get them through finished ceilings or walls, you are going to have to open a channel for them. This will mean repairing the path once the installation is inspected and approved. Again, this is a consideration to keep in mind when you are estimating a job.

Once the water pipes are in the new bathroom location, they will be roughed in by your plumber to accept the fixtures that will be installed (Figure 5.22). Plumbers know what the rough-in measurements for standard fixtures and faucets are. When necessary, plumbers can refer to rough-in guides for specific locations in which to install the piping.

INSPECTIONS

Adding a new bathroom opens you up for the need to have inspections on the job site by local code officers. You must have all plumbing and electrical work inspected before you conceal it. For your own protection, require your subcontractors to provide

TABLE 5.5 Water Distribution System Design Criteria Required Capacities at Fixture Supply Pipe Outlets

FIXTURE SUPPLY OUTLET SERVING	FLOW RATE[a] (gpm)	FLOW PRESSURE (psi)
Bathtub	4	8
Bidet	2	4
Combination fixture	4	8
Dishwasher, residential	2.75	8
Drinking fountain	0.75	8
Laundry tray	4	8
Lavatory	2	8
Shower	3	8
Shower, temperature controlled	3	20
Sillcock, hose bibb	5	8
Sink, residential	2.5	8
Sink, service	3	8
Urinal, valve	15	15
Water closet, blow out, flushometer valve	35	25
Water closet, flushometer tank	1.6	15
Water closet, siphonic, flushometer valve	25	15
Water closet, tank, close coupled	3	8
Water closet, tank, one piece	6	20

For SI: 1 psi = 6.895 kPa, 1 gallon per minute (gpm) = 3.785 L/m.
a. For additional requirements for flow rates and quantities, see Section 604.4.

you with approved inspection slips before you pay them for work performed. Additional inspections will be required once fixtures are set and the job is considered finished. Again, get approved inspection slips before paying your subcontractors.

When you go into bid a major bathroom job, go in with your eyes open. Take trade experts with you when you visit a potential job site. This is the only way that you can make a safe bid. These jobs can be extremely profitable, but they can also eat you alive in costs that you might not anticipate if you do the inspec-

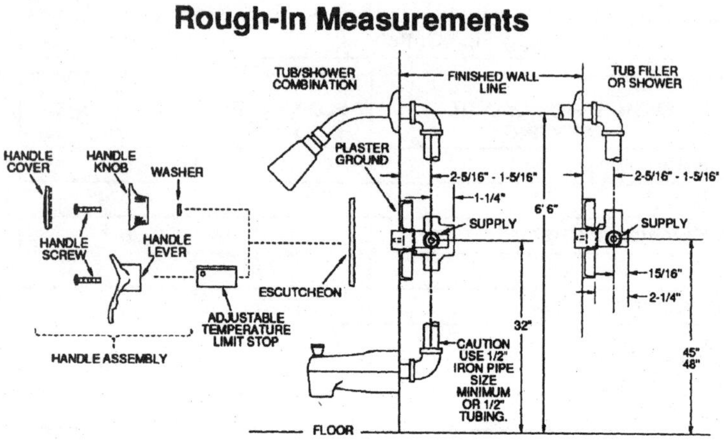

Rough-In Measurements

FIGURE 5.22 A typical rough-in specification for use by plumbers.

tion and estimate on your own. If you are prepared, organized, and professional in your actions, major bathroom remodeling can be your cash cow.

QUESTIONS AND ANSWERS

Q: Does it make sense for me to hire a full-time plumber if I am going to do a lot of bathroom remodeling?

A: *Plumbers are expensive employees. You will need significant volume to justify a payroll plumber. I would use independent contractors until you have enough ongoing volume to justify having a plumber on your payroll.*

Q: I've got a talented carpenter who can do basic plumbing work. Should I allow the carpenter to do the plumbing on my simple remodeling jobs?

A: *You will have to check your local code requirements. Typically, if you are doing basic fixture replacement in the same location, a plumbing permit is not required. It's very possible that your carpenter can handle this type of work. However, confirm that your location allows unlicensed people to perform the plumbing duties that you have in mind.*

Q: How difficult is it to install ceramic tile?

A: *Anyone with average mechanical skills who can follow instructions can set ceramic tile. Some special tools are needed, but they can be rented until you decide if you want to invest in them.*

Q: How can I get rid of old mildew stains on painted walls?

A: *Scrub the mildewed area with a bleach-based cleaner. Apply a stain blocker to the affected area, and then paint the wall.*

Q: I've got a bathroom where old tile runs about halfway up the wall. The customer wants the tile removed and the wall painted. Should I assume that I remove the tile and paint the wall, or should I plan on installing new drywall?

A: *To be safe, plan on installing new drywall.*

Q: Is it true that plumbing must be relocated when a wall-hung lavatory is replaced with a pedestal lavatory?

A: *Yes.*

Chapter 6

ROOM ADDITIONS

Room additions are very much like new construction. The work is considered remodeling, and there are remodeling skills needed, but room additions involve many of the same elements that new construction does. When a room addition is built, it connects to an existing structure. Existing plumbing, heating, and electrical systems are often tapped into for the new addition. There are, of course, times when an addition has new systems due to existing systems being either inaccessible or inadequate.

Much is involved in the planning and creation of a room addition, and the process starts long before construction. If you have experience as a homebuilder, it will serve you well when working with additions. There are plenty of remodeling contractors who never build a room addition. It is not uncommon for remodelers to concentrate all of their efforts inside of existing structures. Other remodelers concentrate on additions as their specialty. There is a lot of money to be made on building additions.

Contractors who work with additions must deal with issues that interior remodelers do not. For example, you will have to check zoning laws to confirm that an addition can be added legally. You will have to confirm the location of the water service and sewer that serves the building. Soil conditions are another factor, since a foundation will be needed. It is truly a from-the-ground-up operation.

SITE VISIT

You will have to conduct a site visit to evaluate the feasibility of adding a room addition to a home. What should you look for? There are many factors to consider. It's a good idea to create a customized checklist to use when performing your site visits.

Example of a Site Visit Checklist

- ✔ Is the proposed location for the addition in compliance with zoning regulations?
- ✔ Will all setback requirements from the property line be met?
- ✔ Are there covenants and restrictions on buildings that must be considered?
- ✔ Does the property use a septic system that must be protected?
- ✔ Where does the sewer leave the property?
- ✔ What is the location of the water service?
- ✔ Is the site accessible by excavation equipment and concrete trucks?
- ✔ What landscaping damage will have to be repaired upon completion of the addition?
- ✔ Will the soil support a suitable foundation?
- ✔ What type of foundation will be needed?
- ✔ Where will the addition connect to the existing structure?
- ✔ What type of roof will be needed?

✔ Will existing mechanical systems be used for the new addition?
✔ How much access is available to existing mechanical systems?

The checklist above is representative of the types of questions you will want answers to before you offer a bid on a job. While this checklist doesn't cover all possibilities, it does provide a good model for you to use in building a customized checklist.

ZONING

Zoning requirements vary greatly from location to location. Before you build an addition, you must confirm that the structure will not violate current zoning requirements. In addition to zoning, you may have to deal with covenants and restrictions. It is not unusual for a community to place covenants and restrictions in property deeds to maintain a uniform look and value. The covenants and restrictions can cover anything from keeping livestock in the backyard to the type of mailbox that may be used. Seriously, you may have to use a particular type of siding or roofing to avoid a lawsuit. Don't build an addition without checking the covenants and restrictions.

You may need to hire a surveying firm to give you a site plan to confirm that the addition will not violate setback laws and regulations. This may seem like a lot of preliminary work and you may want to jump right in and start digging footings, but you had better do the prep work first.

Where can you find out about zoning and covenants and restrictions? Covenants and restrictions are found in the deeds of

Did You Know: that covenants and restrictions are placed in force by developers or land sellers? These are not municipal rules. Zoning laws are the municipal version, but private covenants and restrictions can be much more strict than zoning laws. While zoning may require a minimum setback of 15 feet, private restrictions may limit the setback to 30 feet. You simply have to know what you are facing before you break ground.

property. Ask the homeowner to have a copy of their deed available for your review when you do your site inspection. Zoning requirements can be checked at the local zoning office in your community. Don't skip this step in your planning.

ATTACHMENT

What type of attachment to the existing structure will the new addition require? It is quite common to have an addition attach to an existing home with a lot of distance involved in the attachment. Some additions have their gable end as the point of attachment. Then there are satellite additions where the main portion of the addition is connected to the existing structure with a hallway. The means of attachment has a direct effect on the amount of remodeling work that must be done on the existing structure.

Satellite additions require the least amount of work on the existing structure. But, this type of design is not always appealing to customers. If you have customers who want a long run of the addition to connect to an existing wall, you have to factor in certain potential concerns (Figure 6.1).

Potential Concerns When Connecting An Addition to an Existing Structure.

- ✔ Will existing windows be covered by the new addition?
- ✔ How will access be provided to the addition?
- ✔ What type of roof design will be needed?
- ✔ Are there any existing wall features that will be affected by the addition?

FOUNDATION

The foundation for a room addition is one of the first steps in the actual construction. There are several choices when it comes to foundations. The most cost-effective foundation is a monolithic pour that provides the footing, the foundation, and a concrete floor all in one whack (Figure 6.2).

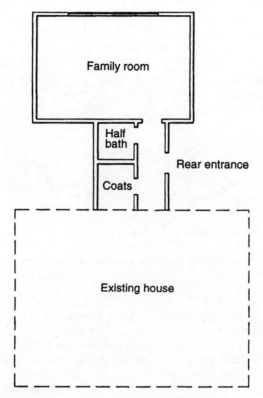

FIGURE 6.1 Example of a satellite addition.

 Trade Tip: Most trench-style footings are about twice as wide as they are thick. Since most footings are 8 inches thick, they are generally 16 inches wide. The width of the footing should be wider in softer soils.

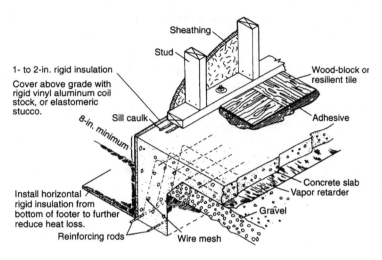

FIGURE 6.2 Combined floor slab and footing foundation system.

Footings are needed for any type of foundation. The footings can be has simple as holes in the ground for a pier foundation. Most footings involve a trench that concrete will be poured into. In any case, the base of the footing has to be on solid ground. This can result in a lot of digging. You must excavate to a depth with solid, undisturbed ground can be found. Additionally, the depth of the footing must be below the local frostline. When I was a builder in Virginia, the minimum footing depth was 16 inches. In Maine, the minimum depth is 48 inches. You will have to check your local requirements to determine the minimum depth of footings in your area.

Soil Loads

- ✔ Soft clay will normally have a bearing capacity of 2,000 pounds per square foot.
- ✔ Loose sand and hard clay can generally support up to 4,000 pounds per square foot.
- ✔ Hard sand or gravel is often rated with a bearing capacity of 6,000 pounds per square foot.
- ✔ Partially cemented sand or gravel might hold up to 20,000 pounds per square foot.

Did You Know: that you can buy foundation vents that open and close automatically based on temperature? You can, and they are a good selection for crawlspace foundations. The vents will close when the air temperature is cold and open when the temperature is warmer. This automatic option is done with the use of a coiled spring.

Crawlspaces

Crawlspaces are an economical type of foundation that allows access beneath the floor structure for the installation of mechanical systems and insulation. When crawlspaces are used, they must be vented. Without the proper venting, the wood structure contained in the crawlspace will suffer (Figure 6.3).

A slab foundation is relatively inexpensive, but many homeowners prefer a wood floor and a raised foundation (Figure 6.4). When this is the case, a crawlspace is a good option. This is also a good way to ensure that the floor level of the addition will be level with the floor of the existing home.

FULL FOUNDATIONS

Full foundations offer many advantages over crawlspaces and slabs (Figure 6.5). At the least, a basement provides generous

	Multiply free vent area by	
Vent cover material	With soil cover	No soil cover
¼" mesh hardware cloth	1.0	10
⅛" mesh screen	1.25	12.5
16-mesh insect screen	2.0	20
Louvers + ¼" hardware cloth	2.0	20
Louvers + ⅛" mesh screen	2.25	22.5
Louvers + 16-mesh screen	3.0	30

FIGURE 6.3 Crawlspace gross vent requirements.

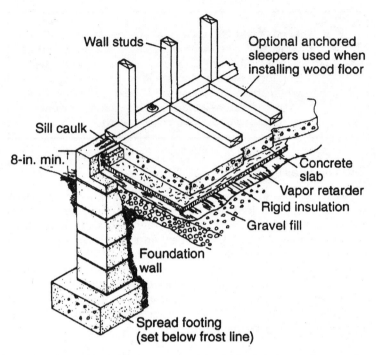

FIGURE 6.4 Independent concrete slab and foundation wall system for deep frostline climates. This can be used when a concrete floor is desired on a raised foundation.

storage. Under enhanced conditions, a basement allows living space with a low cost on a per-square-foot basis. The downside to a full basement is the cost of building it. If bedrock is present, the digging of a basement is not possible, and blasting is rarely an acceptable option on a remodeling job.

Basement walls are often made with poured concrete that is poured into special forms that are removed once the concrete has hardened. A more traditional wall will be made of cinderblock and may or may not have a brick façade (Figure 6.6). Waterproofing material should be applied to any basement wall before it is backfilled. Drain tile, which is slotted plastic pipe, should be installed on a bed of crushed stone and covered with a topping of crushed stone prior to backfilling.

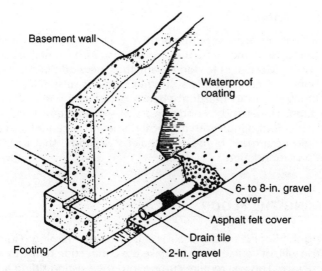

FIGURE 6.5 Poured concrete basement wall keyed to footing.

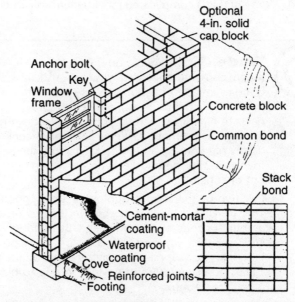

FIGURE 6.6 Block foundation wall.

Pier Foundations

Pier foundations are a very inexpensive way to support an addition, but they come with their problems. Appearance is one major issue. Most people don't want their new addition sitting on a bunch of concrete piers. In cold climates, it is difficult to keep plumbing from freezing when it is installed within a pier foundation. Under most conditions, I would advise against the use of a pier foundation with a typical home. An exception to this rule could be a rustic home or a chalet-style home that already uses a pier foundation.

FRAMING THE FLOOR

Framing the floor is no big deal. This is just basic framing (Figure 6.7). You will, of course, have to make the connection to the existing structure. This will require cutting into the existing siding and flashing the connection. But any experienced carpenter can handle this. Then the sill sealer goes down and the sill plates are installed. Band boards are next. Then you are ready for beams and joists. None of this is complicated for experienced carpenters.

Trade Tip: When you need to know how much concrete to order, you can usually get a good estimate from your concrete suppliers. If you measure the depth, width, and length of what you need to fill with concrete, your suppliers can give you a solid estimate. Order a little more concrete than what you expect to need. There is little worse than being in the middle of a pour when you run out of concrete.

Get two or three price quotes for your concrete. This will give you two or three estimates on the amount of concrete you will need. By figuring your concrete this way, you shift a lot of the responsibility to your supplier. It's better than doing the calculations yourself and having no one to blame if you are wrong.

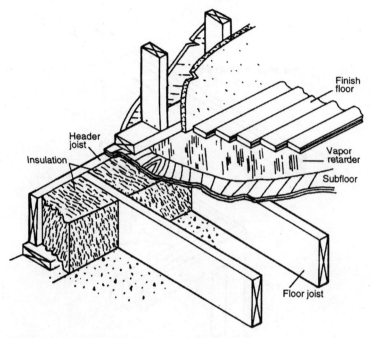

FIGURE 6.7 Typical detail of floor framing.

Check your local code requirements for joist spacing. It is standard procedure to space the joists on 16 inch centers. Blueprints should confirm the size and spacing of joists and beams. If you don't have detailed plans and specifications, you should consult your local code book for the requirements as they pertain to the type of addition you are building (Figures 6.8 through 6.12).

Contractors should have solid plans and specifications to work with. They don't always spend the money for them. Most code enforcement offices require structural plans to be filed for approval before a permit is issued. When this is the case, all you have to do is follow the approved plans and specs. Many remodelers don't have blueprints drawn for small jobs, but a room addition should always have a confirmed set of plans and specifications.

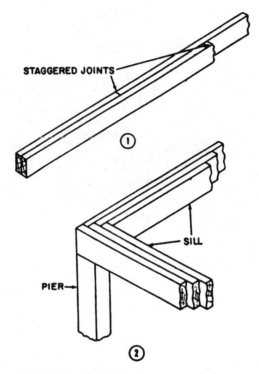

FIGURE 6.8 Sill fabrication.

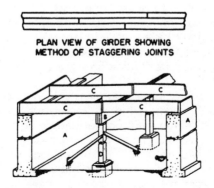

FIGURE 6.9 Built-up girder.

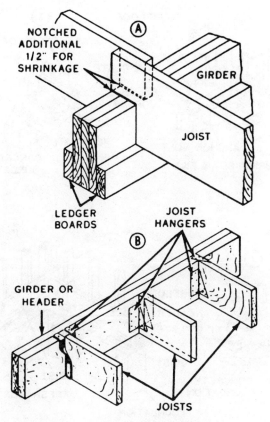

FIGURE 6.10 Joist-to-girder attachment.

 Trade Tip: As a general rule of thumb, 2 x 8 joists will normally be adequate for a simple addition. This is only a rule-of-thumb calculation. You must check your local code book for specific requirements. If you use 2 x 10 joists, you can be almost certain that you will be okay, but this could be a bit of overkill.

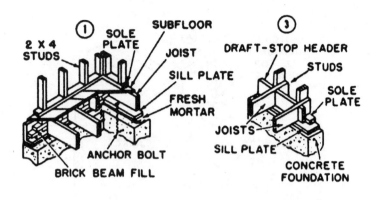
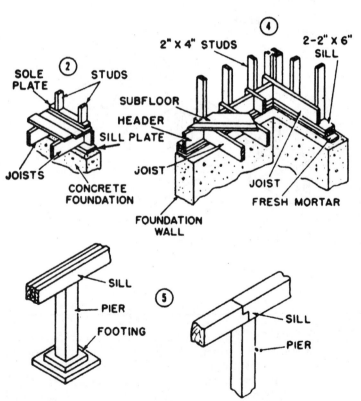

FIGURE 6.11 Types of sills.

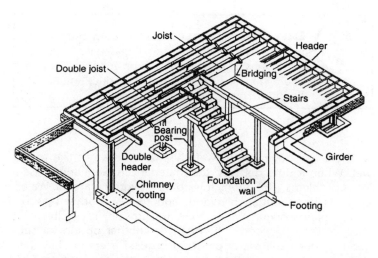

FIGURE 6.12 Framing for a floor system when a basement is installed.

 Don't Do This! You may be tempted to go with the smallest floor joists that you can, but don't do it! If you think you are squeaking by with 2 x 6 joists, upgrade them to 2 x 8s. It will cost a bit more in lumber, but it will be a bargain when compared to what the results of underbuilding could be.

FLOOR SHEATHING

Floor sheathing is usually plywood or waferboard. There are still carpenters who use boards to build a base floor, but they are few and far between. Most building codes require two layers of subflooring, unless tongue-and-groove (T&G) material is used. When T&G is used, one layer is usually sufficient. My preference is ¾ inch, T&G plywood. This is a one-step subflooring material that installs quickly.

I grew up in a time when CDX plywood was the typical subflooring. It is still a favored material, but waferboard is less expensive and frequently used.

Did You Know: that modern procedures rely more heavily on adhesives than nails for securing floor sheathing? The reason for this is that if you eliminate nails, you greatly reduce squeaks in flooring. This can be very effective.

There is a downside to waferboard. It does not like to get wet. When it does get wet, the material can swell and cause a lot of problems. This is why it is not my favorite material. Waferboard is heavier than plywood, and this is another reason I don't enjoy working with it. On the upside, it's quite a bit cheaper than plywood. Fiberboard is another option for subflooring, but I still prefer plywood. I guess everyone has their own preference, but I do lean heavily in the direction to T&G plywood for speed and durability.

WALL FRAMING

Wall framing for an addition is no more complicated than framing any other type of walls (Figures 6.13–6.17). In fact, it can be a lot easier than most framing done in remodeling. Since additions are similar to new construction, you have better working conditions than what are normally experienced with interior framing.

Standard Steps in Framing Walls

✔ Arrange the longest outside wall.
✔ Cut all your framing members.
✔ Nail the members together to create the wall section.
✔ Raise the wall once it is nailed together.
✔ Install braces to hold the wall in place.
✔ Arrange your next wall section.
✔ Repeat the erection stage and bracing.
✔ Join the walls together.
✔ Erect all exterior walls first.

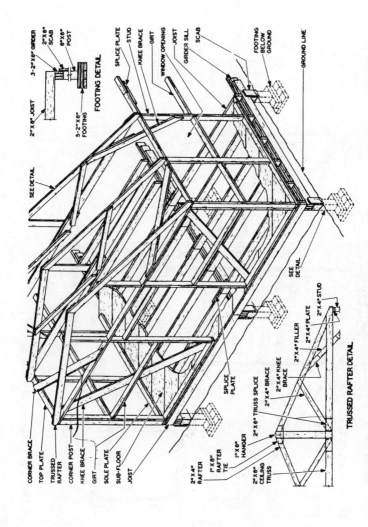

FIGURE 6.13 Example of a rough framing detail.

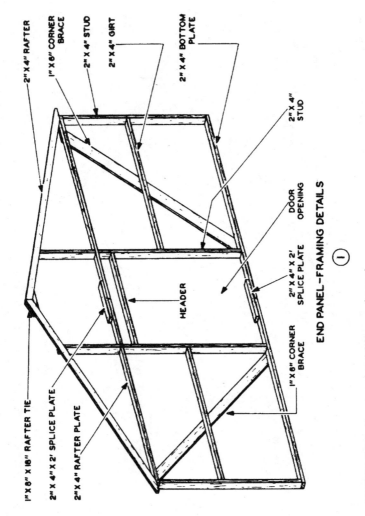

FIGURE 6.14 Framing detail for an end panel.

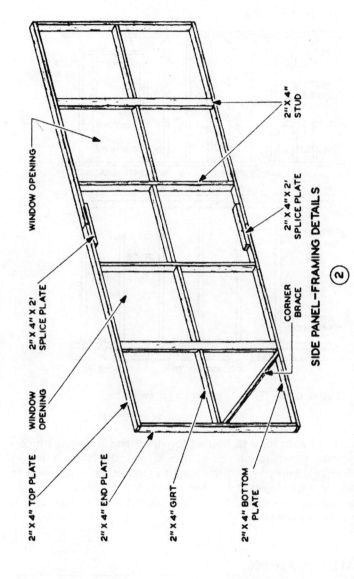

FIGURE 6.15 Typical wall panel.

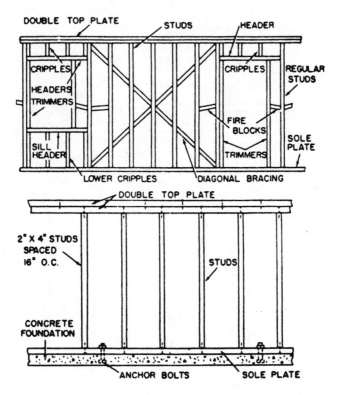

FIGURE 6.16 Typical wall framing details.

Framed walls have to be tied together. This is done with corner-post construction and T-post construction (Figures 6.18–6.22). Bridging is often used to strengthen a wall and to reduce the spreading of fire. Most bridging in walls is done with a horizontal installation. Diagonal bridging is more common in the bays of floor joists.

WALL SHEATHING

Wall sheathing ranges from composite materials to plywood. Most remodelers use plywood, or waferboard on the exterior corners and either rigid foam insulation boards or other composite boards

Room Additions

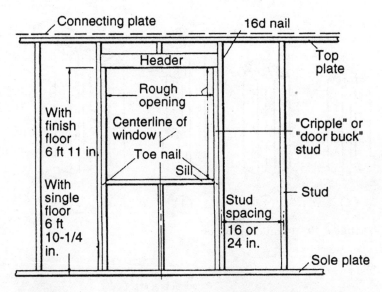

FIGURE 6.17 Typical framing with a header for a window opening.

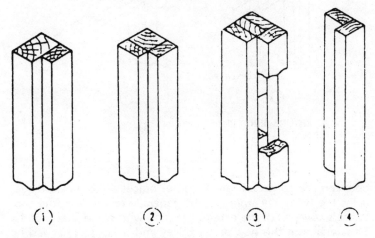

FIGURE 6.18 Corner post construction.

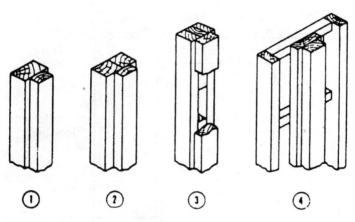

FIGURE 6.19 T-post construction.

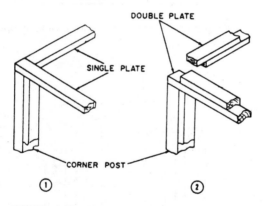

FIGURE 6.20 Plate construction.

between the corners. If metal corner braces are installed diagonally, the wood sheathing is not generally required. Some contractors cover all of the exterior framing with wood sheets of 4 x 8 material, but this gets pretty pricey and really isn't needed for most applications. There is no mystery here for experienced carpenters. You install the sheathing, cut out the rough openings for windows and doors, and move on.

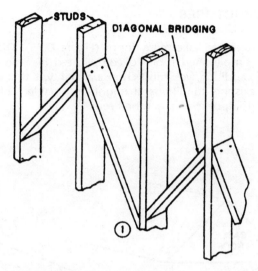

FIGURE 6.21 Horizontal bridging.

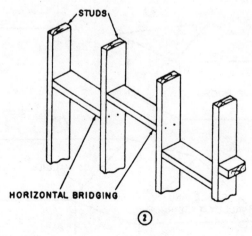

FIGURE 6.22 Diagonal bridging.

ROOF STRUCTURES

Roof structures can take many forms (Figure 6.23). They range from very simple shed roofs to complicated hip roofs. When building an addition, you must choose a roof type that will work well with existing conditions. It might be a gable roof, a shed roof, a hip roof, or even a gambrel roof.

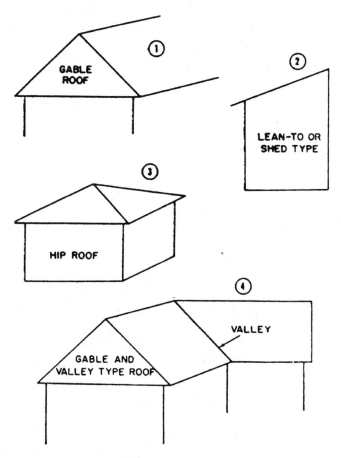

FIGURE 6.23 Types of roofs.

Trusses

Trusses are very common in modern construction (Figures 6.24, 6.25). They are certainly a timesaver. Stick-built roofs remain popular and are frequently used. Trusses tend to be lighter in weight, faster to install, and structurally sound. You pay a little more for the advantages over what you would pay for lumber to frame your own roof, but the savings and benefits are substantial if you factor in all elements of the roof structure.

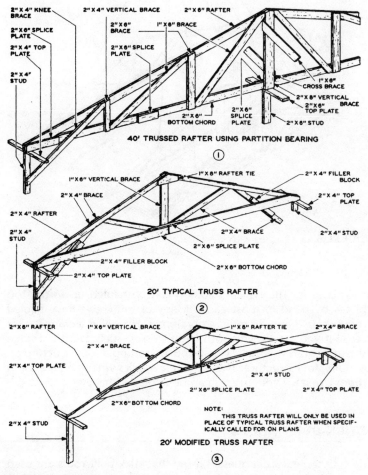

FIGURE 6.24 Truss details.

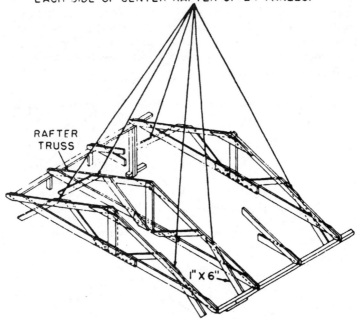

FIGURE 6.25 Trusses with knee braces.

Trusses for the typical addition can be hauled up to the top plates by hand in most cases. Many of my crews have hauled them up by hand for multi-story homes. However, a crane makes sense if you are dealing with large trusses, such as attic-storage trusses. This is especially true if you are setting a roof structure on a multi-story structure. The same rule applies when you are using room trusses. These trusses can be large and heavy.

STICK-BUILT ROOFS

If I had to venture a guess, I'd say that stick-built roofs are used more often on room additions than trusses are (Figure 6.26).

Did You Know: that you can order room trusses that provide you with a pre-framed room in the attic space? You can, and they work pretty well. This is something to keep in mind for some room additions. You can buy lumber and do a lot of framing, or you can set trusses and have the attic rooms already framed for you.

There is nothing wrong with a stick-built roof, but it can be a waste of time and money when trusses will get the job done. Some homeowners want nothing other than a stick-built roof. Other homeowners don't know the advantages of using trusses. It's your job to consult with your customers and inform them of what is most likely to be best for them.

Many framing techniques can be applied when building a roof structure (Figure 6.27). There is very little that you can do in traditional roof construction that cannot be accomplished with a truss system. It probably sounds like I am a big fan of trusses, and I am. But, I still use a lot of stick-built roofs. You have to match the roof structure to the job and please your customers.

Stick building a roof does allow a carpenter ultimate freedom to overcome on-site problems. This is something that cannot be overlooked. Trusses are delivered as they are, and they are what they are. There is little flexibility in what can be done with them if an on-site modification is needed. Personally, this is one of the factors I weigh when deciding whether to use trusses or raw lumber to create a roof structure.

Remodeling is a tricky business. Rarely do things go as you plan them. When this happens, it is helpful to have raw lumber to work with. I believe that you have to evaluate every job on its own merit to decide what type of roof structure is the most appropriate.

Most trusses and rafters are set 24 inches on center. Determining rafter size is done with your local code book. You must factor in dead loads on a regional basis. Roof pitch is also a factor in determining what size lumber is needed. It stands to reason that a roof with a 6/12 pitch is going to need stronger lumber in a region where heavy snow is expected than a roof with a 10/12 pitch would need. The sheer weight of the load is the factor.

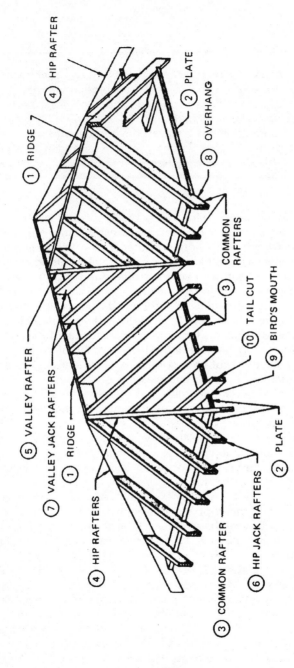

FIGURE 6.26 Detail of a stick-built roof.

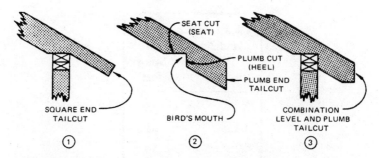

FIGURE 6.27 Common cuts on bottoms and ends of rafters.

Sizing lumber for a roof structure is critical, whether you are building a roof or setting trusses. Trusses tend to be engineered with smaller lumber than stick-built roofs. Take a look at Figures 6.28, 6.29, and 6.30 for what the spans are for different types of trusses. Compare this with your local code book for a stick-built roof. How many 2 x 4 rafters have you seen used in modern building construction? Not many.

Chord size: 2-×-4 top and 2-×-4 bottom

Pitch	Span (feet)
2/12	22
3/12	29
4/12	33
5/12	35
6/12	37

FIGURE 6.28 Typical truss spans (55 PSF with 15% duration factor).

Chord size: 2-×-6 top and 2-×-4 bottom

Pitch	Span (feet)
2/12	28
3/12	39
4/12	46
5/12	53
6/12	57

FIGURE 6.29 Typical truss spans (47 PSF with 33% duration factor).

Chord size: 2-×-6 top and 2-×-6 bottom	
Pitch	Span (feet)
2/12	32
3/12	51
4/12	56
5/12	60
6/12	62

FIGURE 6.30 Monopitch truss span (55 PSF with 33 percent duration factor).

Deciding on what type of roof structure to use is affected greatly by existing conditions. If you are inclined to use trusses, make sure that you won't be buying yourself trouble. I like trusses, but sometimes remodelers are better off to build their own roof structures.

ROOF SHEATHING

Roof sheathing is a simple proposition. The most common roof sheathing is either CDX plywood or waferboard. Some contractors use T&G plywood to cover a roof structure. A more common approach is to use plain CDX plywood and plywood clips. The clips support the edges of plywood between spans. Wood boards are occasionally used to build the roof decking (Figure 6.31). This is a fairly rare practice today. I do see some of it here in Maine, but 4 x 8 sheets of plywood or waferboard are much more common.

The material you choose for roof sheathing will affect the allowable live load on a roof. For example, if you install ½ inch plywood with a maximum span of 32 inches and a maximum unsupported edge length of 28 inches, the allowable live load, when rafters are set on 24 inch centers, will be 95 pounds per

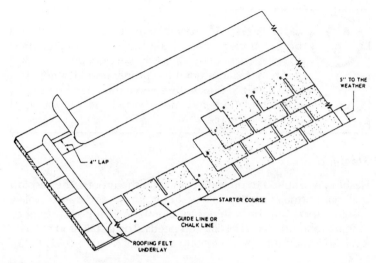

FIGURE 6.31 Detail of a board roof.

square foot. Installing ¾ inch plywood with a maximum span of 42 inches and an unsupported maximum edge length of 32 inches, the live-load rating will be 145 pounds per square foot. Check your local code book for more examples of load ratings. Once you decide on the type of sheathing that you will use and know what your load rating requirements are, installing the sheathing goes quickly.

ROOF COVERINGS

Roof coverings are available in many styles. Asphalt shingles are the primary choice for residential roofs. These are usually three-tab, standard shingles. Dimensional asphalt shingles are frequently used for a more defined look. In either case, these shingles require an underlayment that is usually created with roll felt roofing material. A 15 pound felt is standard procedure. The felt should be installed as soon as the roof decking is complete. This phase is often referred to as drying-in a building. Keeping the roof deck dry is important. If it gets wet and warps, it will have a negative affect on the finished roof's appearance.

Did You Know: that failure to install a proper underlayment can void the warranty on shingles that you install! Many manufacturers will not honor warranties if the proper underlayment is not installed. Check the shingle manufacturer's recommendations for underlayment before you install their shingles.

Flashing

Flashing is required to prevent leaks in finished roofs. The flashing material normally used is rolled aluminum. In addition to valley flashing, there may be flashings for plumbing vent pipes, chimney flashing, and other special flashing needs. The installation of flashing is an important step in creating a dry roof (Figures 6.32, 6.33).

Ventilation

Ventilation is critical for a healthy attic and roof. Soffit venting is the most common means of roof ventilation. Many contractors use ridge vents in conjunction with soffit vents. When this is done, special sheets of foam boards with pre-formed channels in them are installed in the attic between rafters or trusses. The channels in these foam boards allow air to move freely from the soffit vent to the ridge vent.

Turbine vents have been used for attic ventilation, and they work well. However, many people don't like the large, visible turbines adorning their roofs. Ridge vents are much less obtrusive and generally more readily accepted as a good means of ventilation.

Ice Dams

Ice dams can do a lot of damage to a building when the structure is located in a cold climate. Living in Maine, I am well aware of the risks of ice dams. There are three basic methods used to reduce or eliminate damage from ice dams.

When I am doing a roof in Maine, I use a membrane material that runs from the edge of the roof to a point several feet up on the roof. This waterproof membrane greatly reduces the risk of water backing up under shingles and destroying the roof structure.

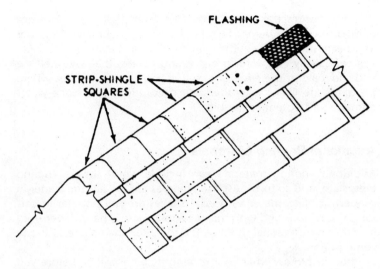

FIGURE 6.32 Open valley flashing.

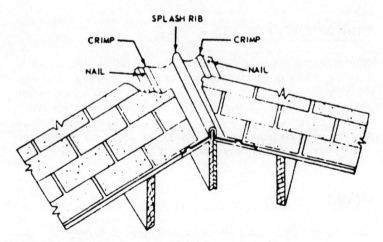

FIGURE 6.33 Hip or ridge flashing.

Some contractors use a wide aluminum flashing to prevent damage from ice dams. This method is also effective. Another approach that is sometimes used is the installation of a heat source, usually in the form of heating cables, that will melt ice and snow along the edge of a roof. If you work in an area where ice dams are likely, take adequate precautions to reduce water damage and potential lawsuits.

Additional Roofing Materials

Additional roofing materials can include wood shakes, roofing tiles, slate, and so forth. Asphalt shingles are, by far, the leader in modern roofing materials. Cedar shakes are still popular in certain subdivisions and with many people. Slate is hardly ever used in modern, residential roofing. Tile roofs see their share of use in some regions.

You must know what your roofing material will be before you frame a roof. The weight of the roof is a serious factor. Obviously, a slate roof will weigh much more than an asphalt roof will. Don't frame a roof until you know what you will be covering it with.

WINDOWS AND DOORS

Windows and doors are installed prior to siding. The basics of installing these units are nothing fancy. If the rough openings were framed correctly, it doesn't take long to install windows and doors (Figure 6.34). General carpentry experience is all that is required to set windows and doors in place. It is, however, critical that the units be installed properly. Once siding is installed, correcting errors with windows and doors becomes a major issue.

SIDING

The type of siding installed on a room addition is normally consistent with the siding that is on the existing structure. Many types and styles of siding are possible (Figures 6.35, 6.36). My personal favorite is beveled wood siding. Vinyl siding is very popular with many people. Hardboard siding can give an elegant look at affordable expense when the siding will be painted, but

Room Additions

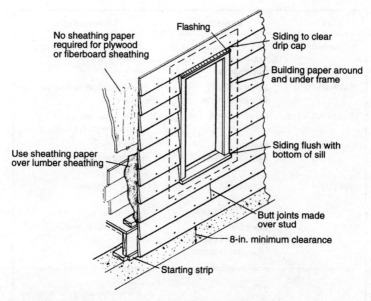

FIGURE 6.34 Detail of siding installed around a window.

Material	Care	Life, yr	Cost
Aluminum	None	30	Medium
Hardboard	Paint Stain	30	Low
Horizontal wood	Paint Stain None	50+	Medium to high
Plywood	Paint Stain	20	Low
Shingles	Stain None	50+	High
Stucco	None	50+	Low to medium
Vertical wood	Paint Stain None	50+	Medium
Vinyl	None	30	Low

FIGURE 6.35 Siding comparison.

Material	Advantages	Disadvantages
Aluminum	Ease of installation over existing sidings Fire resistant	Susceptibility to denting, ratting in wind
Hardboard	Low cost Fast installation	Susceptibility to moisture in some
Horizontal wood	Good looks if of high quality	Slow installation Moisture/paint problems
Plywood	Low cost Fast installation	Short life Susceptibility to moisture in some
Shingles	Good looks Long life Low maintenance	Slow installation
Stucco	Long life Good looks in SW Low maintenance	Susceptibility to moisture
Vertical wood	Fast installation	Barn look if not of highest quality Moisture/paint problems
Vinyl	Low cost Ease of installation over existing siding	Fading of bright colors No fire resistance

FIGURE 6.36 Siding advantages and disadvantages.

hardboard siding is not usually considered an option when stain is preferred over paint. It's rare to see aluminum siding used in new construction. Shake siding is another option, but it is time consuming to install and tends to be expensive. When wood siding is installed, it is most often installed horizontally. However, vertical wood siding is another option.

Vinyl Siding

Vinyl siding is extremely popular among people who want the most maintenance-free siding that they can get for their homes. Some people shun vinyl siding as a cheap alternative to wood siding. When it comes to personal preference, it's difficult to pinpoint what an individual will want. Vinyl siding does offer an economical solution to siding a home, and the material is easy to clean and maintain. Durability is good with this type of siding. Many colors are available and there are even wood-grain patterns for people who like the wood look but want the vinyl advantages.

Shake Siding

Shake siding doesn't show up on a lot of new houses these days. This type of siding tends to be expensive, time consuming to install, and can be expensive to maintain. Some houses are still fitted with shake siding, but I can't recall doing a job like this in the last decade, or more.

Hardboard Siding

Hardboard siding is used frequently with certain styles of homes. This is an inexpensive siding that can give an elegant look, especially when it is painted. The framing on the structure must be plumb to make this siding look good. Hardboard siding conforms to the surface that it is applied to. If the rough framing is not straight, you will be able to see lumps and curves in the siding. Check your framing carefully before using hardboard siding (Figures 6.37, 6.38).

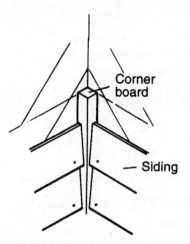

FIGURE 6.37 Inside corner board in use.

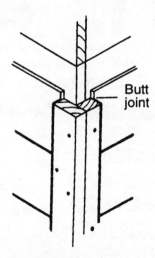

FIGURE 6.38 Outside corner board in use.

Wood Siding

Wood siding is often the siding of choice (Figures 6.39, 6.40). I have found that my customers prefer beveled wood siding over all other types of siding. As an individual, I agree with the choice. Cedar is the high end of this type of siding. Pine is a cheaper alternative. Both woods can be stained, but pine tends to have more blemishes in it. This is fine on rustic homes. Some people insist on clear cedar, which is expensive. Whether painted or stained, wood siding is hard to beat for appearance. Unlike vinyl siding, wood does require periodic painting or staining. All in all, I think wood is still king in the hearts of homeowners.

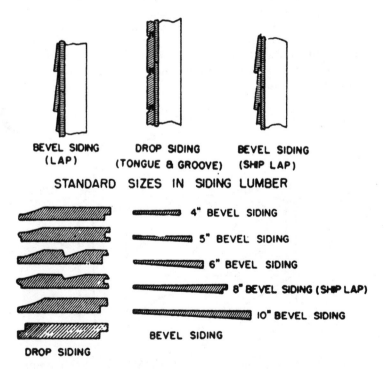

FIGURE 6.39 Types of wood siding.

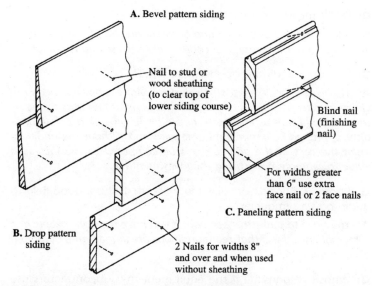

FIGURE 6.40 Methods of nailing wood siding.

MOVING INSIDE

Once you have the addition framed and dried in, you will be moving inside to finish the space. Much of the interior work is covered in previous chapters. If you need help on specifics that are not in previous chapters, check Chapter 13 for details.

QUESTIONS AND ANSWERS

Q: Is there good money in building room additions?

A: *There is plenty of profit potential in room additions.*

Q: Is it suitable to use a monolithic pour for a slab floor in a room addition?

A: *This is largely a question for your customers. Slabs are used often in many parts of the country. On the whole, they are frequently considered to be of less value than a foundation and a wood floor.*

Q: Don't most room additions have shed roofs?

A: *Shed roofs are quite common on room additions, but the deciding factors are the customers' choices and the existing conditions.*

Q: Are room additions good investments?

A: *This is a difficult question to answer. Many room additions are viable investments. If the rate of return is of great importance, a customer should hire a real estate appraiser to provide a before-and-after appraised value for the home with and without the addition. Then the cost of the addition can be compared to the increased value to estimate how strong the investment is.*

Q: Room additions seem like a lot of work. Is this a good field to get into as a remodeler?

A: *You have to make this decision. If you have the right people and some patience, I feel that room additions are a good area of remodeling to get into.*

Q: Who is responsible if the equipment my subcontractors use damage the lawn while building a room addition?

A: *I am not an attorney, but my guess is that the general contractor will be responsible. This is a good issue to address in the contract with your customer.*

Chapter 7

SUNROOMS

Sunrooms are a very popular home improvement. They are incredibly popular in most regions. Most of my work has been done in Virginia and Maine. The two states are very different in climates, but sunrooms are extremely popular in both states. There can be some design differences between various climates, but the idea of having a sunroom plays well in any climate.

Many of the sunrooms in Maine don't incorporate extensive ventilation. This is in contrast to sunrooms in southern states where ventilation is often a major factor. Solarium-style glass panels are popular in colder climates while casement windows and sliding-glass doors are fashionable in warmer climate.

When it comes to style and design, sunrooms open a window of creativity. They can be passive-solar rooms, glorified screened porches where glass is added, and so forth. The cost for sunrooms can range from moderate to very expensive. Our goal in this chapter is to discuss construction and remodeling options when creating a sunroom.

ADDING A SUNROOM

Adding a sunroom to a home is similar to building a room addition. You can refer to the previous chapter for many of the details on what to look for when adding space to a home. Zoning requirements, covenants, restrictions, foundation matters and similar steps will be essentially the same with a sunroom as they would be with a room addition. There are differences in the framing process, due to the increased amount of windows, glass, and doors. Roofing is often the same, but solarium sunrooms use panels that create a glass roof, so traditional roofing is not needed. Let's concentrate on the differences between sunrooms and typical room additions. To do this, we will explore some key elements on a one-by-one basis.

Zoning

Zoning, covenants, and restrictions for the addition of a sunroom are very similar to room additions. Essentially, you can follow the guidelines in Chapter 6 on this issue. As you were told in that chapter, don't scrimp on your research when it comes to the legality of adding to the footprint of an existing structure.

Foundation Work

Foundation work for a sunroom falls under the same basic procedures discussed in Chapter 6 for room additions. There is basically no difference between a room addition and a sunroom when it comes to foundation requirements.

Framing the Floor Structure

Framing the floor structure for a sunroom is done with the same procedures used for room additions and new construction. Again, there is nothing special here to talk about.

FRAMING THE WALLS

Framing the walls of a sunroom can require different tactics from what you would use for a room addition. For example, if you are using one-piece solarium panels that make up the wall sections and ceiling, you will have a very different framing program. Many sunrooms incorporate the use of sliding-glass doors, patio doors, casement windows, double-hung windows, fixed glass, or other visual options. When these items are used, the framing is more conventional, except that you have many more headers and rough openings (Figure 7.1).

In wood framing, you will have to connect wall sections. This is done like any traditional framing. Partition posts are used to accomplish this (Figure 7.2). You can use any type of connection arrangement that you prefer.

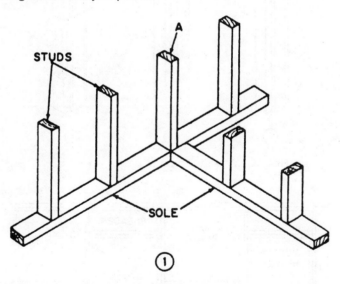

FIGURE 7.1 A typical framing diagram that shows the sole plates and studs.

 Trade Tip: Sunrooms offer a lot of profit potential. If you are not involved in the creation of sunrooms, you probably should revisit your business plan.

Doors and windows will be roughed-in the same as they would be for any other type of new construction (Figure 7.3). This involves the use of headers, jack studs, cripple studs, and typical routine framing.

You can use any acceptable framing procedures that feels comfortable to you and acceptable to your customers. This can mean joining walls with different methods, framing rough openings in your own way, or sticking with traditional methods (Figures 7.4, 7.5). As long as your way is in compliance with local codes, you should be fine.

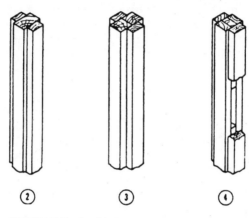

FIGURE 7.2 Partition posts.

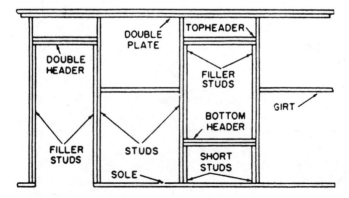

FIGURE 7.3 Door and window framing.

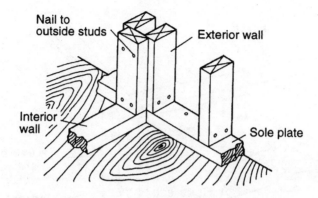

FIGURE 7.4 Tying an interior partition to an exterior partition.

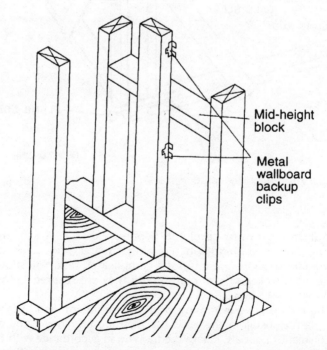

FIGURE 7.5 Using blocking as an attachment means for connecting an interior partition to an exterior partition.

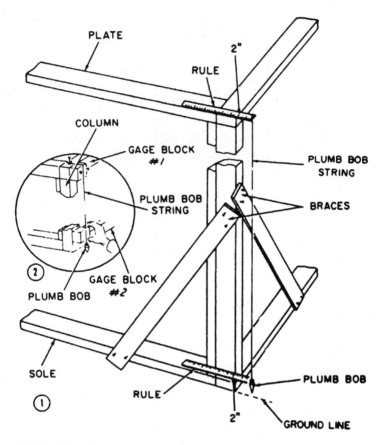

FIGURE 7.6 Plumbing a post.

Wall sections must be plumb (Figures 7.6, 7.7). This is true of any quality framing job. Framing squares and levels are needed for this job, but it is not a complicated task. You should brace walls to keep them plumb until they are secured in their final form.

The nature of your framing work will be determined by what you plan to install in the walls that you are framing. If you are using solarium panels, create the framing in accordance with the manufacturer's recommendations that are provided with the solarium panels.

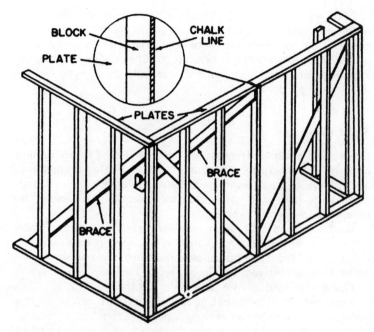

FIGURE 7.7 Straightening a wall.

If you are dealing with typical windows and doors, use the same rough-in measurements that you would use for any other type of framing. In all cases, make sure that the rough openings are sized properly. A small mistake in the framing stage will result in a major headache as you move further through the job.

ROOFING

Generally roofing for a sunroom is the same as it would be for a room addition. It is likely that you will be installing skylights or roof windows that you might not install in a room addition. This is not a big deal, but it does require a little special framing work on the roof structure.

Roof windows and skylights must be framed with the proper rough opening. This is usually as simple as nailing a few headers between rafters. Some units are curb units that require a raised platform for mounting the skylight to. Check your rough openings and mounting requirements carefully before you call the roof framing a done deal.

Roof Windows

I love roof windows. They are my favorite way of bringing natural light in from a roof. There are many reputable brands available. Some of them even have blinds built into them. I don't particularly favor this style, due to potential problems if the blinds fail to operate properly.

Roof windows that open and that have screens are great. Depending on the location, these windows can be opened by hand, by pole, or by a power pole that turns the window mechanism to open the roof window.

Good roof windows are expensive, but they are not so expensive that they must be ruled out even in sunrooms of modest cost. Ventilating roof windows and skylights offer much more versatility than fixed skylights do. If your customer can swing it, I strongly recommend the installation of ventilating units in the roof.

Bubble Skylights

Bubble skylights come in many sizes and are inexpensive when compared to ventilating units. Many bubble skylights are curb-mount units. They are little more than formed panels that are nailed or screwed to a raised curb on the roof. These units are available in a clear or smoked finish. Most of my customers prefer the smoked version. The drawback to a fixed skylight is the lack of ventilation, but the cost is considerably less than it would be for a ventilating skylight.

SOLARIUM PANELS

At first look, solarium panels seem cost prohibitive. They are expensive, but you must take into account what you are getting

Trade Tip: When you install a roof window or skylight, make sure that it is flashed properly. Seal all connection points carefully. Take your time and inspect the work closely. A leak after a job is done will cut into your profit and hurt your reputation.

for your money. A standard sunroom will have siding, windows, and roofing. Solarium panels replace the need for siding, windows, and roofing. If you factor in all the costs, both for labor and material, solarium panels can be a bargain. At the least, they can be a viable option.

There are limitations to what you can reasonably do with solarium panels. Primarily, you are limited in the size of the sunroom. The panels come in various sizes, but you don't have the creativity in size and shape of a sunroom made with solarium panels that you can have with a more traditional approach. This is something to keep in mind.

Many franchise dealers specialize in solarium-style sunrooms. It appears that they do quite well. Does this mean that you can't compete with them? Of course not. If you have a good reputation, do good work, and hustle, you can compete on most any course. I certainly would not rule out this type of sunroom application.

SLIDING DOORS

Sliding doors are often used to create fairly simple sunrooms. This is a common practice. The drawback here is that ventilation is limited. Sliding doors only have half of their width available for ventilation. But, they are a cost-effective means for making a sunroom on a tight budget (Figure 7.8).

Building rough openings for sliding doors and setting them is not difficult work. Numerous sunrooms are comprised mostly of sliding doors. The prices for these doors can range from the very inexpensive to the quite expensive. I recommend staying away from the cheap versions, but doors in the middle range will do well.

Glass size (in inches)	Frame size (width × height, feet and inches)	Rough opening (width × height, feet and inches)
33 × 76¾	6-0 × 6-10¾	6-0½ × 6-11¼
45 × 76¾	8-0 × 6-10¾	8-0½ × 6-11¼
57 × 76¾	10-0 × 6-10¾	10-0½ × 6-11¼
33 × 76¾	9-0 × 6-10¾	8-0½ × 6-11¼
45 × 76¾	12-0 × 6-10¾	12-0½ × 6-11¼
57 × 76¾	15-0 × 6-10¾	15-0½ × 6-11¼
33 × 76¾	11-11 × 6-10¾	11-11½ × 6-11¼
45 × 76¾	15-11 × 6-10¾	15-11½ × 6-11¼
57 × 76¾	19-11 × 6-10¾	19-11½ × 6-11¼

FIGURE 7.8 Measurements for sliding-glass doors.

A common problem with sunrooms that have sliding doors or solarium panels is the lack of wall space to rough in electrical outlets. This is more of a problem with solarium panels than it is with sliding doors, but if you don't provide a framed partial wall, it does make wiring more difficult. Floor outlets are one solution to this problem.

Drawbacks to Sliding Doors

- ✔ The tracks can freeze in cold weather when cheap doors are used.
- ✔ Only half of the glass space can ventilate the sunroom.
- ✔ People think of burglary being easy when sliding doors are installed.
- ✔ Fogging between the glass is not uncommon is cheap sliding doors.
- ✔ Cheap sliders can be difficult to slide in their tracks.

Advantages of Sliding Doors

- ✔ Sliding doors, even those of good quality, are a fair value for a sunroom.

Did You Know: that electrical outlets installed in a floor must be of a waterproof type? This is generally the case, so check you local electrical code to see what is required in your area.

- ✔ Framing is simple and installation is fast.
- ✔ High-quality sliders move easily in their tracks.
- ✔ Sliding doors provide a lot of glass at a reasonable price.
- ✔ Unlike fixed glass, sliders do offer reasonable ventilation.
- ✔ Limited wall space is available when sliding doors are used.

FIXED GLASS

Fixed glass offers a very economical means of creating a sunroom. I have used fixed glass in a multitude of applications to create desirable results. The glass can be colored, textured, or even have designs made into it. Fixed glass is much less expensive then most window options. The disadvantage is that the panels cannot be opened for ventilation. This same type of problem comes with solarium panels. Some form of ventilation is usually essential in a sunroom.

Fixed glass can be used in conjunction with sliding doors, ventilating windows and skylights, and roof windows to round out a sunroom. Etched glass that has theme figures in it can be very popular. I would not overlook this option when bidding a job. Most glass shops are happy to bid the production of fixed-glass panels.

WINDOWS

Windows are a key to sunrooms. You can install fixed glass or solarium panels, but windows are extremely common in sunrooms. Even sunrooms that use fixed glass and sliding doors can house windows. The very nature of a sunroom calls for a lot of light, and windows are a viable source for this asset (Figures 7.9–7.11). The following are types of windows to consider:

- Double-hung windows
- Single-hung windows
- Vertical casement windows
- Horizontal casement windows
- Awning windows
- Horizontal sliding windows
- Vertical sliding windows

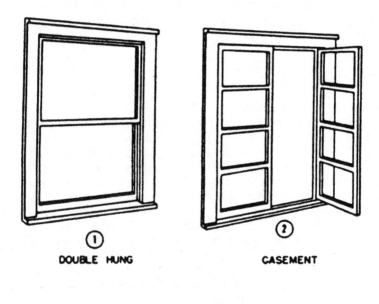
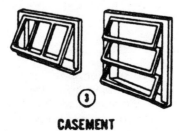

FIGURE 7.9 Examples of window types.

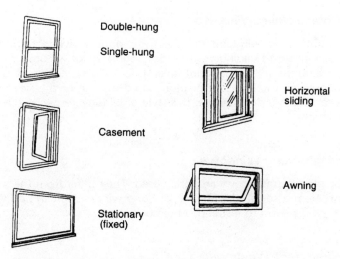

FIGURE 7.10 Common window styles.

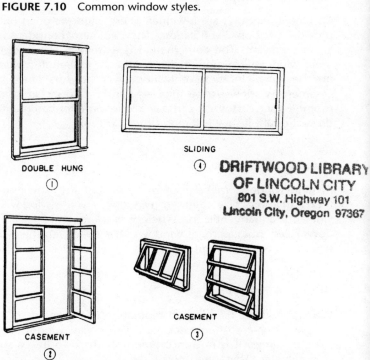

FIGURE 7.11 Types of windows used in sunrooms.

Double-Hung Windows

Double-hung windows are the most popular style of window used in most building construction. This type of window is okay for a sunroom, but it is not ideal. Like a sliding door, a double-hung window offers ventilation based on half of the window's size. This is a drawback to this style of window, especially in hot climates.

Single-Hung Windows

Single-hung windows are rarely used. They suffer from the same problem associated with double-hung windows. I can't think of a good reason to install this type of window.

Casement Windows

Casement windows are a winner. These windows open out and provide additional ventilation. They are also considered very energy efficient. The combination is extremely effective. Casement windows are more expensive than double-hung windows, but they offer a lot more ventilation. Over the years, I have found casement windows to be the best windows to install in a sunroom. If your customers can pay a little more for casement windows, I doubt if they will regret their decision.

Awning Windows

Awning windows make a lot of sense, but they are not used much anymore. Like casement windows, awning windows provide substantial ventilation. There was a time when these windows ruled, but casement windows have overtaken them. Still, I would not rule them out.

Sliding Windows

Sliding windows have a bad reputation in the minds of many people. They are often associated with cheap windows in aluminum frames. It is not uncommon for them to freeze shut in cold climates. There are good sliders available, just as there are good sliding doors. Even so, the ventilation is only equal to half of the window mass. I strongly recommend casement windows for sunrooms.

WINDOW FRAMING AND INSTALLATION

Window framing and installation varies with different types of windows (Figures 7.12, 7.13). There are many types of windows available. You must check the rough-in dimensions for the specific window units you plan to install. This is an essential step. Don't get caught in the trap of doing what you assume is right. Make sure that your assumption is right.

Some windows are set in rough openings and screwed into place. These windows typically have a mounting flange. There are many types of windows in the industry, so you have to be sure of what you are working with (Figure 7.14).

How will you group windows in a sunroom? Will you bank them together, or will you space them out? This is a decision that you should discuss with your customer (Figure 7.15). It's very common to bank the windows in a sunroom, but this is not always the case. My experience shows that windows are generally placed closely together, but you should draw a plan and have your customer sign off on it before you frame the walls for a sunroom.

The basic framing procedures for a sunroom are the same as they are with any other window installation (Figure 7.16). You will have a headers, some spacers blocks, cripple studs, and jack studs. It's no big deal.

Your methods for rough framing will be dictated by the type of window that you will be using. This is not a difficult procedure. Check the rough openings provided by the window manufacturer and adhere to them. Don't guess at what to do. Follow the recommendations to the letter (Figures 7.17–7.20).

 Don't Do This! It can be tempting to offer customers inexpensive doors and windows to win a bid. This is usually a mistake. If you do it, disclaimer it with the potential problems associated with the cheap materials. Your reputation is at stake. I suggest that you offer only products that you are willing to put your name on.

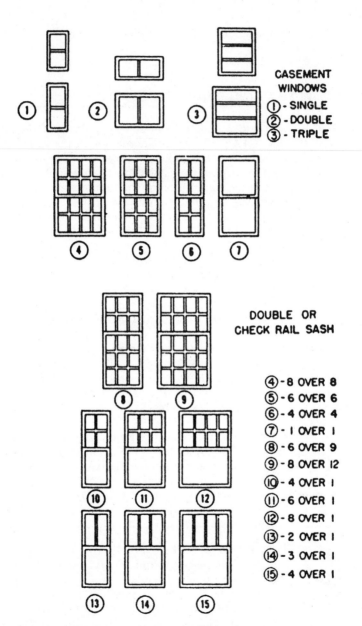

FIGURE 7.12 Types and sizes of milled sashes.

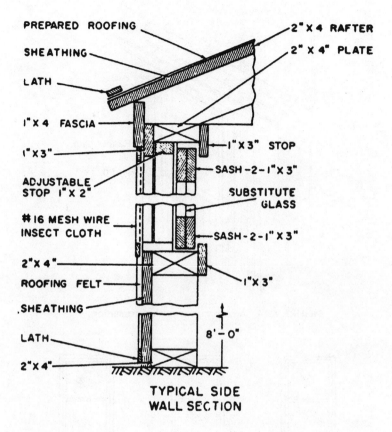
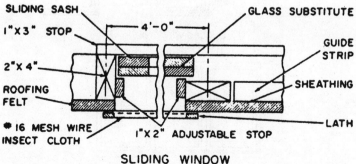

FIGURE 7.13 Sliding window sash installation.

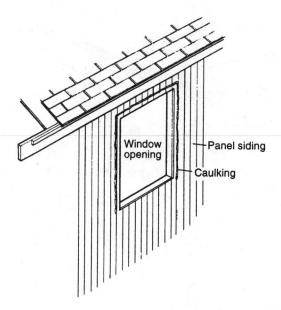

FIGURE 7.14 A common window installation.

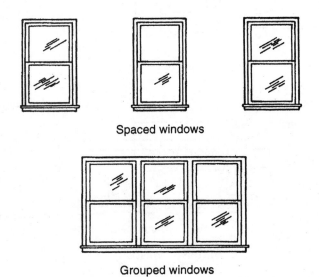

FIGURE 7.15 Window-grouping options.

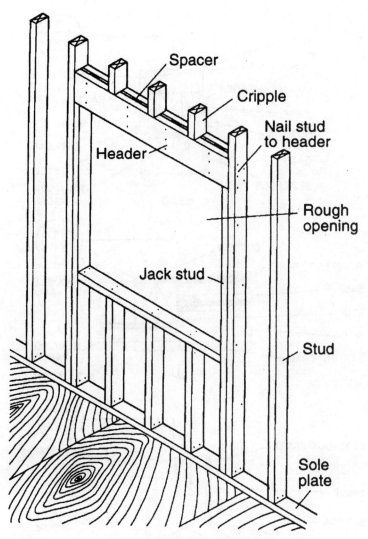

FIGURE 7.16 Framing detail for a typical window.

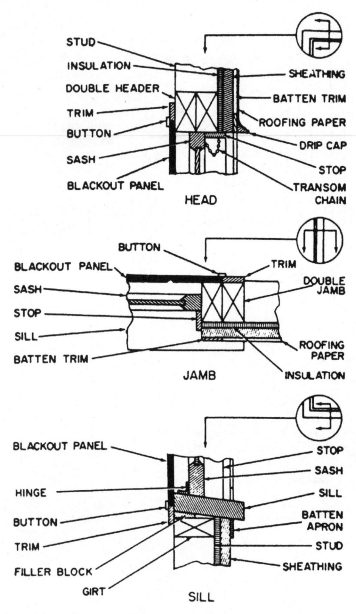

FIGURE 7.17 Detail of wall section with window frame and sash.

Sunrooms

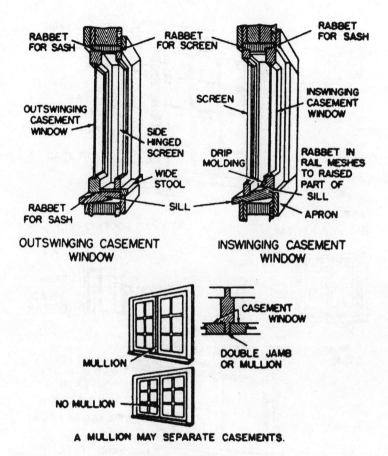

FIGURE 7.18 Casement window details.

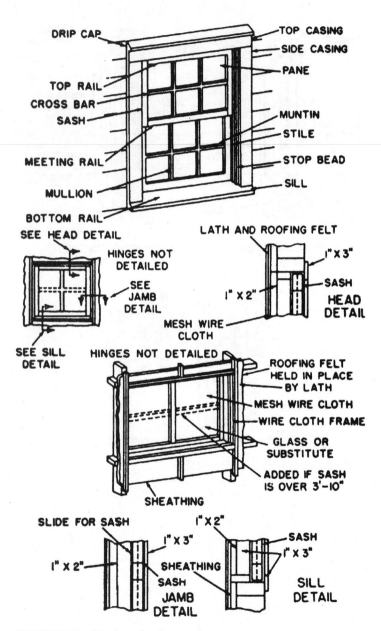

FIGURE 7.19 Window details.

Sunrooms

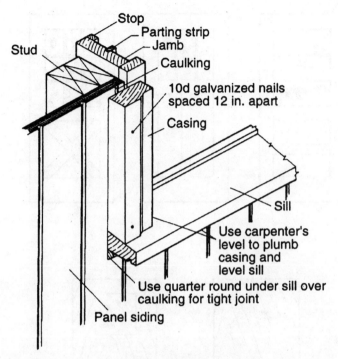

FIGURE 7.20 Installation of a double-hung window frame.

DOORS

Exterior doors in sunrooms are not always needed. However, they are often installed. When they are, they can take several forms. Sliding-glass doors are very common. Patio doors are frequently chosen. French doors are not uncommon. Standard entry doors are also used (Figure 7.21). Most of the doors contain glass panels.

Any type of door used will require a sill (Figure 7.22). When you are dealing with sunrooms, you could use any type of door. I would say that sliding doors are the most common. French doors and patio doors are my personal preference, but they are more expensive and not used as often as sliding doors. Flat doors that don't contain glass are rarely used in sunrooms.

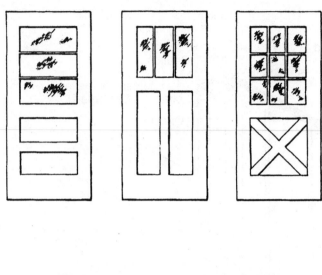

FIGURE 7.21 Types of entry doors.

The ultimate decision on a door choice is in the hands of the homeowner. Personally, I like patio doors. They tend to be less expensive than French doors and more dependable than sliding doors. This is just my personal thought. There are many great sliders that you can use. High-end customers may opt for French doors. The choice is really up to the customer. Your job is to rough out the opening to the proper specs, which is no big deal.

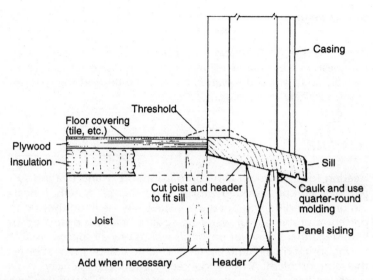

FIGURE 7.22 Door installation at the sill.

ELECTRICAL WORK

Electrical work in a sunroom is nothing special. But it you reduce wall space for running electrical wiring, you make the job more difficult. Assuming that you have adequate access under the sunroom, the wiring can be done effectively. Electricians can run their wires through the attic, if one exists. Many sunrooms have vaulted ceilings. This can restrict electrical options. However, most electrical needs can be met with a vaulted ceiling. Solarium panels present the most problems for electricians when underfloor installations are not possible.

Wire molding can be use to run wiring in exposed areas, but this is an issue that should be addressed up front with your customers. Baseboard molding can be used to create a camouflaged wiring program, but the wiring must be protected. Generally speaking, electrical wiring should be installed from beneath the floor if there is not a partial wall to put the wiring in. Talk to your electrician and find out what your options are.

PLUMBING

Plumbing is not a big factor in most sunrooms. There might be a request for a bar sink. The installation of a hose connection may be wanted to aid in the watering of plants. Beyond this, plumbing is not likely in a sunroom.

HEATING

Heating for a sunroom can come from an existing heating system or it can be provided with an independent system. If heating or air conditioning ducts are installed in the floor, they don't present a problem. Hot-water baseboard heat can pose a problem, since there may not be long sections of wall space available for mounting the heating element. Electric baseboard heat poses the same problem. A wall-mounted electric heater is okay.

Any in-floor heating or cooling system should work fine in a sunroom. The wall where a sunroom adjoins an existing structure offers the opportunity for a wall-mounted heating unit. If baseboard heat is desired, you must make sure that you create enough wall space to accept it.

FLOORING

Flooring is covered in Chapter 3. Sunrooms can be fitted with nearly any type of floor covering. Many customers prefer carpeting. I like quarry tile. You will have to discuss flooring options with your customers to determine what is wanted. Your options are open with a sunroom.

 Don't Do This! Contractors sometimes skimp on heating a sunroom. They assume that the room will be solar heated. Passive solar heating will occur, but it should not be considered a substitute for an adequate heating plan. If you are working in an area where cold temperatures are common, don't skimp on heating a sunroom.

ACCESSORIES

Accessories for sunrooms often play a large part in the successful completion of the job. Ceiling fans are very common in sunrooms. False ceiling beams are commonly used in sunrooms where traditional or vaulted ceilings are installed. Custom blinds are another accessory found in sunrooms. Plant shelves are another common enhancement to sunrooms. Don't overlook the opportunities that exist in putting accessories in sunrooms. It can lead to more money in your pocket and happier customers.

QUESTIONS AND ANSWERS

Q: Are those kits that you see for sunrooms any good?

A: *That's a broad question and difficult to answer. Some kits are good, and others are not so good. You will have to do your homework and make your own evaluation.*

Q: Does the brand of casement window I install really make a difference?

A: *It certainly can. There are times when people pay a lot for a name brand and don't get enough additional value to offset the increased cost. But, there are plenty of cheap windows out there that will make you regret using them. I recommend sticking with known windows that have a good track record among contractors.*

Q: Wouldn't it make sense to use fixed-glass panels to create a sunroom?

A: *This is a cost-effective means of creating a sunroom, but it comes with a problem. Since the glass will not open for ventilation, excessive heat is a potential problem. I believe in using fixed glass in conjunction with either windows or sliding doors that do open.*

Q: Can I sell sunrooms as being passive solar rooms?

A: *I would be very careful about this, unless you incorporate all elements of a passive solar structure. The fact that a room is built with glass walls, in my opinion, does not qualify as a solar room. Again, I am not an attorney and can't give you a legal decision, but I wouldn't make the passive solar angle my major sales pitch.*

Q: I've heard that cheap sliding doors are a mistake when it comes to professional use. Do you agree with this?

A: *I don't recommend budget sliding doors at all. They don't glide well and they often fog up.*

Q: How do you feel about patio doors for sunrooms?

A: *I like patio doors when they are used to get from a home into a sunroom, but I don't favor them as the primary glass in a sunroom. Since the doors have to swing to open, they limit furniture placement.*

Chapter 8

ATTIC CONVERSIONS

Attic conversions can be tricky business. The process makes a lot of sense. For homeowners, the return on their investment is usually better with attic conversions than it is with basement conversions. Remodeling an attic to accommodate living space requires a lot of thought and planning. The process can be reasonably simple or extremely complicated. Don't think for a moment that this type of work is simple or easy.

Many factors come into play during an attic conversion. The degree of complexity depends largely on existing conditions and the intended changes. Here are just some of the things you must think about:

- Will the attic floor support the weight requirements of the conversion?
- How will natural light be provided to the attic?
- What means of ingress will you create?

- Will the roof slope allow a ceiling height that will comply with code requirements?
- What are the risks of damaging living space below when remodeling the attic?
- Will the attic require plumbing?
- Can the existing electrical service support the new circuits required?
- How will the attic space be heated or cooled?
- Do existing stairs provide suitable access or will new stairs be needed?

The above list is only a portion of what you will have to consider when asked to convert an attic to living space. If there are existing stairs, suitable framing, and no need for plumbing, the conversion of an attic can go pretty smoothly. More often than not, you will encounter many problems during the remodeling of an attic.

Attic conversions are easier than room additions in that you are not extending the foundation size of the building. That's about where the easy part stops. If you are not experienced in attic remodeling, you will probably take a beating from time to time on your estimates. I can tell you from experience that this type of work can get nasty fast.

TYPE OF ATTIC CONVERSION

The type of attic conversion that you are asked to do will have a lot to do with the nature of the work. Some people convert an attic into a large playroom for their children. Another common tactic is to turn an attic into a bedroom. Both of these types of conversions are fairly easy. If you are asked to put multiple bedrooms and a bathroom in an attic, the process becomes more complicated. When you have to build a kitchen into the space, to make the equivalent of an independent apartment for children or relatives, you up the ante on difficulty.

Since attics are above finished living space, getting mechanical systems to the attic area is not an easy task. You may have to choose an area below to use as a chase to get plumbing, heating, cooling, and electrical connections to the attic. This is often a closet that a homeowner allows you to compromise. If you can't use a chase area, you may have to open up an existing wall

Trade Tip: Create a company procedure for the way that you evaluate attic jobs. Use a customized checklist to make sure that you don't overlook important elements, such as stairways.

and run the mechanical equipment through the wall. Repairing the wall and repainting the room is an expense that you must factor into your estimate. When you estimate an attic job, make certain that you have a defined plan to work with.

STAIRS

Stairs are a major consideration when it comes to an attic conversion. Some attics are equipped with existing stairs that can be used for living space. Other attics have stairs for storage use of the attic. These stairs might not meet code requirements for living space, so check them closely. You might find an attic that has no stairs or that has a pull-down staircase. These are the tough jobs. Making stairs work where there are none can be very difficult.

It is not the framing of the stairs that is so difficult. Your problem will come in finding a place to put the stairs. This is not always a problem, but it can be. At the very least, you have to factor the location of stairs into the equation of how hard it will be to do an attic conversion.

Where will you install stairs for an attic conversion? If you are dealing with a one-story house, the options improve. It's very challenging to find a way to get stairs from an attic to living space in a multi-story home. Attic-storage stairs are frequently located in bedrooms. These can be used, but few people want traffic to and from the attic going through a bedroom.

Hallways in multi-story homes are not an ideal place for a full-sized staircase. Will you run the stairs down the outside of the home? This is an option, but it is not one that many people like. An exception to this could come into play when the attic is used as a child's private apartment. If you plan to use an outside staircase, you have to check the zoning laws, setbacks, covenants, and restrictions that may apply to the property. Once you know where the stairs will fit, you have to plan their construction. This is not particularly difficult for seasoned carpenters, but it is a job that must be thought out (Figures 8.1–8.3).

Trade Tip: Spiral staircases are a natural choice when space is limited and access to an attic is needed. As good of an idea as this seems, it comes with problems. Moving furniture up a spiral staircase is not fun. Try getting a king-size mattress up tiny stairs and see how much fun you have. If you suggest spiral stairs, make sure that your customers are aware of the restrictions that they may face by using the spiral system.

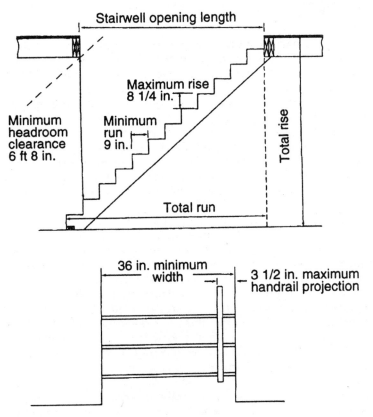

FIGURE 8.1 Stairway design requirements.

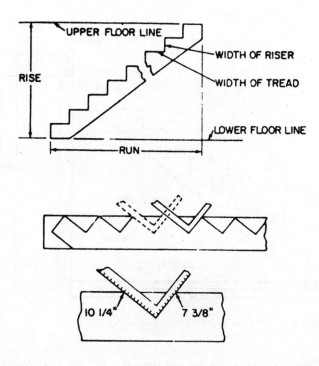

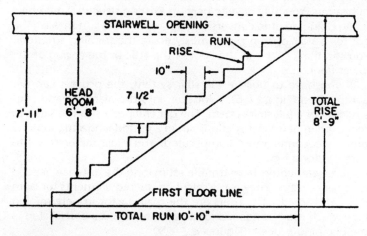

FIGURE 8.2 Principal parts of stair construction.

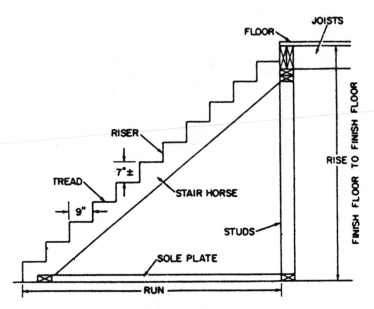

FIGURE 8.3 Method of laying out stair stringers.

When existing stairs are not present, you will have to create an opening in the ceiling for them. This requires the installation of headers. You may have to beef up existing structural members when doing an attic conversion (Figures 8.4–8.5). In making the cut for a stair opening, you must secure the framing structure. You can think of this as the floor of the attic or the ceiling of the lower floor.

If you have to build a full stair system, the process can get complicated (Figure 8.6). A lot goes into the planning and construction of a full stair system. The general process is the same as that used in new construction, but in a remodeling situation, the process becomes more complicated. You must factor this into your bid for a job.

Some carpenters have trouble estimating the distance needed for a set of stairs. They must calculate the requirements for both risers and treads. The risers are the main factor in determining how much room is needed for a stairway. This is a crucial step in the planning process (Figures 8.7–8.9).

A straight stairway is easy to frame, but it can be difficult to work into a remodeling situation (Figure 8.10). You may be lucky

Attic Conversions

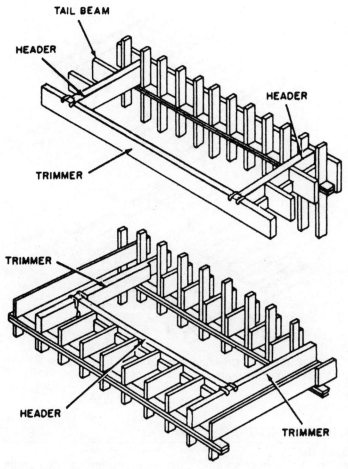

FIGURE 8.4 Heading off a floor opening for an attic.

enough to be working with a house where a straight staircase is already installed for access to an attic for storage purposes. This simplifies the process considerably.

Spiral stairs are great when they suit the customers' needs (Figure 8.11). Keep in mind that moving furniture up spiral stairs is difficult. If your customers are comfortable with spiral stairs, they are a space-saving means of providing access to attic space.

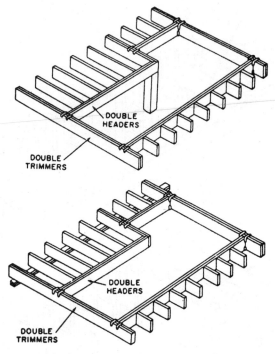

FIGURE 8.5 Doubling headers and trimmers for additional strength.

FLOOR STRUCTURE

The floor structure of an attic must be strong enough to handle the new load that an attic conversion will put on it. The system doesn't have to be very strong to stored stuffed animals, holiday decorations, and other lightweight belongings. However, converting the attic space to living space requires a very different evaluation.

How an attic will be used affects the structural requirements of the floor structure (Figures 8.12–8.15). Live-load ratings affect the size and spacing of floor joists. Many attics are framed with joists or truss components that will not support the load requirements of living space. I have seen attics where the joists were 2 x 4s. Obviously, a 2 x 4 joist is not going to support much weight.

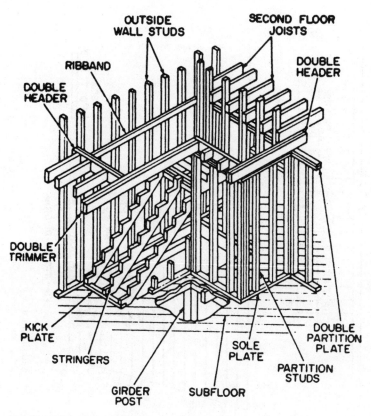

FIGURE 8.6 Details of a complete stair system.

When this is the case, you have to install new, larger joists. Keep in mind that doing this will reduce ceiling height, and this can present its own problem.

Installing larger joists in an attic is not difficult. The old joists don't have to be removed. New joists can be installed beside the existing joists. Since the joist will sit on the plates, the ceiling below doesn't have to be damaged or disturbed. However, there is frequently damaged caused to ceilings below an attic being remodeled. Nails pop out, cracks occur, globes of light fixtures fall, and other accidents happen. Secure the lower living space as well as you can before you start banging around a lot in an attic.

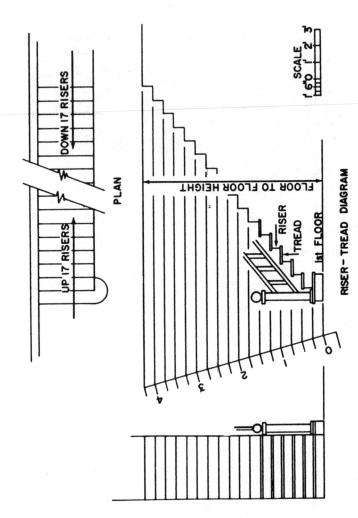

FIGURE 8.7 Details of riser-tread diagram.

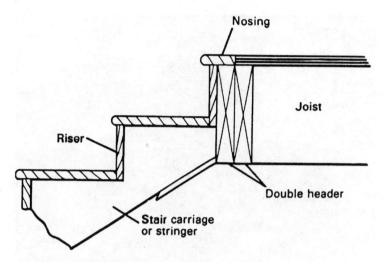

FIGURE 8.8 Typical stair riser diagram.

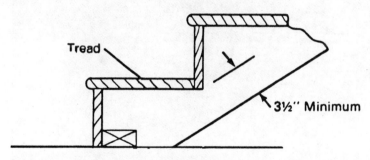

FIGURE 8.9 Typical stair tread diagram.

 Don't Do This! It is tempting to use existing joists that are close to large enough to support the needed loads for an attic conversion. Don't take chances on undersized or borderline joists. Upgrade the joists to ensure a safe floor structure. Whatever money and time you save in cutting a corner will likely come back to haunt you at far more expense.

FIGURE 8.10 Straight stairway.

Basic framing fundamentals should be used when building or strengthening an attic floor structure (Figures 8.16–8.18). You don't want new framing to sag and damage a ceiling below the attic floor. When you are attaching joists to a beam, you should install either a ledger strip or use joist hangers to ensure that the joists will not move out of position.

FIGURE 8.11 Spiral stairs.

Bridging is used to keep joists from twisting and creating nail pops and cracks. Modern construction often uses metal strips for bridging. This is a simple, cost-effective means of bridging. Some carpenters prefer solid bridging. Other carpenters like wood bridging used in a crossing fashion. The type of bridging used is not as important as the fact that it is used (Figure 8.19).

Material	Load per PSF
Standard 2 × 4 (16" on center)	2
Standard 2 × 6 (16" on center)	2
Standard 2 × 8 (16" on center)	3
Standard 2 × 10 (16" on center)	3
Softwood, per inch of thickness	3
Hardwood, per inch of thickness	4
Plywood, per inch of thickness	3
Concrete, per inch of thickness	12
Stone, per inch of thickness	13
Carpet, per inch of thickness	0.5

FIGURE 8.12 Dead load weights of various flooring materials.

- The first floor of a residential dwelling should be rated at 40 pounds per square foot (PSF).
- Other floors should be rated at 30 PSF.
- Stair treads should be rated at 75 PSF.
- Roofs used for sun decks should be rated at 30 PSF.
- Garages for passenger cars should be rated at 50 PSF.
- Attics accessible by stairs or ladder should be rated at 30 PSF, when the ceiling height is more than 4½ feet.
- Attics accessible by scuttle hole with a ceiling height of less than 4½ feet can be rated at 20 PSF.

FIGURE 8.13 Minimum uniformly distributed live loads.

Area/activity	Live load, PSF
First floor	40
Second floor and habitable attics	30
Balconies, fire escapes, and stairs	100
Garages	50

FIGURE 8.14 Residential live loads.

Joist size (inches)	Spacing (inches)	Pine/fir (feet/inches)
2 × 6	12	10-6
2 × 6	16	9-8
2 × 6	24	8-4
2 × 8	12	14-4
2 × 8	16	13-0
2 × 8	24	10-4
2 × 10	12	17-4
2 × 10	16	16-2
2 × 10	24	14-6
2 × 12	12	20-0
2 × 12	16	18-8
2 × 12	24	16-10

FIGURE 8.15 Typical maximum floor joist spans with a 40 pound live load.

Since it is difficult, to say the least, to nail bridging to the bottom of a joist in an attic conversion, solid bridging is popular. The solid bridging can be nailed easily without disrupting existing ceiling structures.

Plumbing Vents

Plumbing vents usually pass through an attic on their way to open air. The pipes can present problems for remodelers. If you can arrange a partition to sit over the vent pipe, you don't have a problem. When this won't work, your plumbing contractor will have to relocate the vent. Plumbing codes frown on 90 degree turns in main vents. A 45 degree offset is preferred over a 90 degree bend. This is not always possible, and most plumbing inspectors will work with you under these circumstances. However, you cannot count on getting a variance. If you see a vent pipe that will be a problem, bring it to your customer's attention before you take the job (Figure 8.20). Have the customer sign off on your warning.

	Length of maximum clear span (ft,in.) for lumber with various MOE ($\times 10^6$ lb/in^2)										
	1.0	1.1	1.2	1.3	1.4	1.5	1.6	1.7	1.8	1.9	2.0
	Living area (40 lb/ft^2 live load)										
Minimum required bending stress (lb/in^2)	920	980	1,040	1,090	1,150	1,200	1,250	1,310	1,360	1,410	1,460
Joist size											
2 by 6	8,4	8,7	8,10	9,1	9,4	9,6	9,9	9,11	10,2	10,4	10,6
2 by 8	11,0	11,4	11,8	12,0	12,3	12,7	12,10	13,1	13,4	13,7	13,10
2 by 10	14,0	14,6	14,11	15,3	15,8	16,0	16,5	16,9	17,0	17,4	17,8
2 by 12	17,0	17,7	18,1	18,7	19,1	19,6	19,11	20,4	21,9	21,1	21,6
	Sleeping area (30 lb/ft^2 live load)										
Minimum required bending stress (lb/in^2)	890	950	1,000	1,060	1,110	1,160	1,220	1,270	1,320	1,410	1,360
Joist size											
2 by 6	9,2	9,6	9,9	10,0	10,3	10,6	10,9	10,11	11,2	11,4	11,7
2 by 8	12,1	12,6	12,10	13,2	13,6	13,10	14,2	14,5	14,8	15,0	15,3
2 by 10	15,5	15,11	16,5	16,10	17,3	17,8	18,0	18,5	18,9	19,1	19,5
2 by 12	18,9	19,4	19,11	20,6	21,0	21,6	21,11	22,5	22,10	23,3	23,7

FIGURE 8.16 Samples of span distances for floor joists that are installed on 16 inch centers.

Attic Conversions **231**

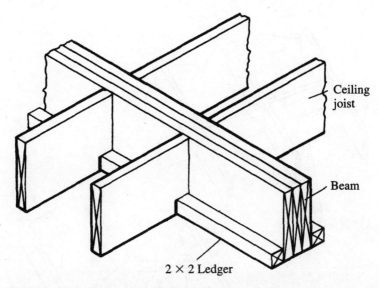

FIGURE 8.17 The use of a ledger strip where joists connect with a beam.

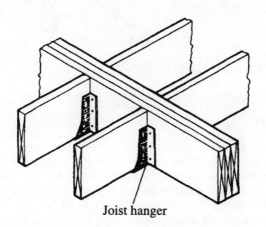

FIGURE 8.18 Detail of joist hangers being used to connect joists to a beam.

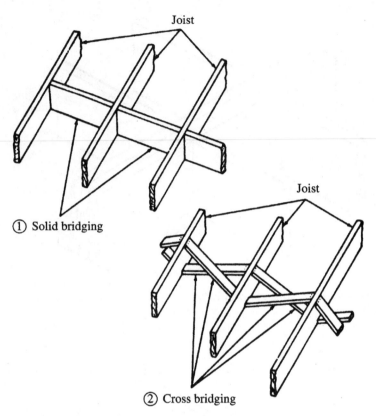

FIGURE 8.19 Types of joist bridging.

Ductwork

Ductwork is common in some attics. Having a trunkline of ductwork running through the middle of an attic can ruin your plans. Building the floor up to conceal the ductwork will most likely eliminate the needed headroom for the living space. It is often possible to relocate the ductwork to one edge of the attic. If this is done, you can do some creative framing to conceal the ductwork.

Duct branches that are lying on top of the existing joists can be relocated to run in the joist bays. While this is not a major job, it is something that you must factor into your bid for a job. With the exception of trunklines, you can work around most ductwork.

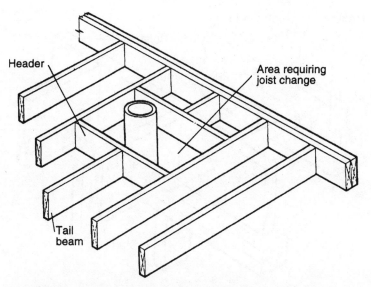

FIGURE 8.20 Detail of framing around a plumbing vent pipe.

PIPING

Some attics contain various types of piping that are often installed on top of existing joists. These can be gas pipes, plumbing pipes, or other types of piping. Relocating the pipes is not usually an insurmountable task (Figure 8.21). Routing the pipes along the edge of a floor will allow the opportunity to conceal the piping favorably.

WIRING

Electrical wiring is often installed all over an attic. In some cases, it is stapled neatly along a board in a single route. Other electricians are not so kind. You may find wires running every which way. This creates a flooring problem. The wires have to be moved or protected.

You can protect the wiring by building a raised flooring platform, if you have enough headroom to work with. The chances are that you will not have the excess headroom to meet the needs. Typically, headroom in an attic is at a premium.

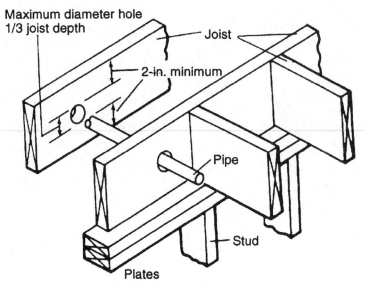

FIGURE 8.21 A detail of installing piping in joists.

ROUGH FLOORING

The rough flooring for an attic is no different than it would be for any other application. There is nothing special here to talk about. If you can deck out a floor structure in a room addition or new construction, you can do it in an attic.

HEADROOM

The amount of headroom required in living space is dictated by local building codes (Figure 8.22). Most ceiling heights are set at 8 feet. It is not uncommon for some rooms to be allowed to have a minimum ceiling height of 7½ feet. What does your local code require? If you don't know, you had better find out. Attic conversions are most often difficult due to a lack of headroom.

Sloped walls are common in converted attics. The primary ceiling height is usually what is at issue. Rafter spacing is another issue to check with your local code requirements. Check your

Don't Do This! You might be tempted to notch the top of joists to conceal electrical wiring. Don't do it. If you cut the top or the bottom of a joist, you compromise its structural integrity. This is not acceptable You will have to either relocate the wiring or build up a flooring surface that will protect it.

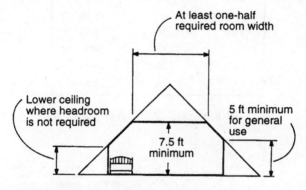

FIGURE 8.22 Headroom requirements.

local code requirements for this measurement (Figure 8.23). If you are running short, moving the collar ties higher might take care of the problem. In other cases, you may have to replace the entire roof structure to comply with code requirements. You can imagine how expensive this can be.

It is common for attic conversions to have flat ceilings, but this is not always the case (Figures 8.24, 8.25). Cathedral or vaulted ceilings are not unusual in any living space, and they are not rare in attic conversions. If roof windows or skylights are used for natural light, a vaulted or cathedral ceiling is ideal. These roof units provide natural light at a much lower cost than a dormer would. Ventilating units enhance the value even more. This is certainly worth running the numbers on.

When roof windows or skylights are used, you must rough-in an opening for them. This is not complicated and we have talked

Rafter size (inches)	Spacing (inches)	Pine (feet/inches)
2 × 4	12	11-6
2 × 4	16	10-6
2 × 4	24	8-10
2 × 6	12	16-10
2 × 6	16	15-8
2 × 6	24	13-4
2 × 8	12	21-2
2 × 8	16	18-10
2 × 8	24	17-104
2 × 10	12	24-0
2 × 10	16	23-8
2 × 10	24	21-4

FIGURE 8.23 Typical maximum spans for roof rafters.

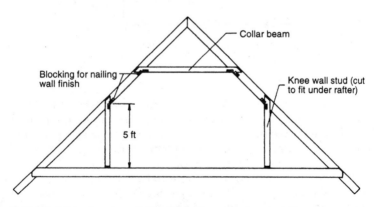

FIGURE 8.24 A conventional flat-ceiling attic framing diagram.

about it previously. You will need proper support and headers (Figure 8.26, 8.27). If you can avoid the need for dormers, you will reduce the overall cost of a job. We will talk more about dormers in the next chapter, but roof windows are a viable option to the more traditional, and more expensive, dormer.

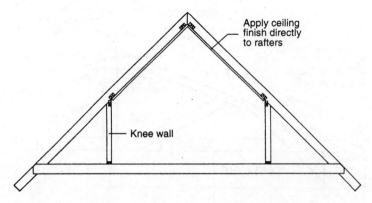

FIGURE 8.25 A sloped ceiling used in a finished attic.

WALLS

Walls installed in an attic conversion are no big deal. The wall studs are nailed to the sill plate on a floor and can be nailed to either a top plate that is attached to the rafters or trusses, or the tops of the studs can be nailed directly to the rafters or studs. You will do better to use a top plate when it comes to hanging drywall. General wall construction for interior partitions is no different than it is for any other type of framing.

WINDOWS

Windows installed in gable ends are framed with standard methods. When windows are needed on a sloped section where roofing is installed, a dormer will be needed. The procedure for framing a dormer is covered in the next chapter.

MECHANICAL INSTALLATIONS

Mechanical installations in an attic conversion are nothing special. They are done with the same procedures that you would use in new construction. The one difference, and it can be a big one,

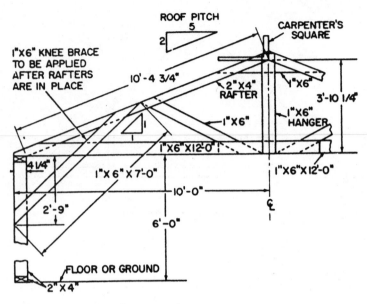

FIGURE 8.26 Roof support details.

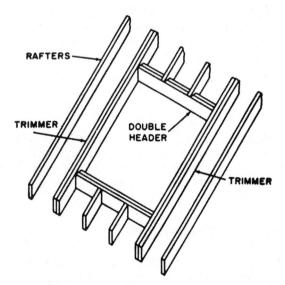

FIGURE 8.27 Framing for a roof window or skylight.

is getting the mechanical materials to the attic. This was discussed earlier in the chapter. Once you get viable plumbing, electrical, heating, and cooling connections to the attic, you can run them normally.

INSULATION

Insulation is always an issue in attic jobs. The first step is dealing with existing insulation. Then there is the need to install new installation for the converted space. Many types of insulation can be found in an attic (Figures 8.28–8.30). Glass-fiber batt insulation is usually the easiest to deal with. This type of insulation is not normally difficult to remove or manipulate. Loose-fill insulation, which is also common in attics, is messy and difficult to move.

Will you remove existing insulation in the floor of the attic? Some contractors do and others don't. Since the attic is being converted to living space, the floor area below heated space is not required to be insulated. Other than for the loss of rising heat from below, there is not much reason to remove old insulation.

Ceiling and exterior wall space in the new attic will need to be insulated (Figures 8.31, 8.32). Batt insulation is almost always used for the wall cavities. The ceiling area can be insulated with either batts or loose-fill insulation.

Thickness (inches)	R-value
3½	11
3⅝	13
6½	19
7	22
9	30
13	38

FIGURE 8.28 Ratings for glass-fiber batt insulation.

Type	R-value per inch of insulation
Cellulose	2.8 to 3.7
Fiberglass	2.2 to 4.0
Perlite	2.8
Rock wool	3.1
Vermiculite	2.2

FIGURE 8.29 Ratings for loose-fill insulation.

Thickness (inches)	R-value
3½	11
3⅝	13
6½	19
7	22

FIGURE 8.30 Ratings for rock wool batt insulation.

Material	R-value per inch of insulation
Fiberglass batts	3
Fiberglass blankets	3.1
Fiberglass loose-fill	3.1 to 3.3 (when poured), 2.8 to 3.8 (when blown)
Rock-wool batts	3
Rock-wool blankets	3
Rock-wool loose-fill	3 to 3.3 (when poured), 2.8 to 3.8 (when blown)
Cellulose loose-fill	3.7 to 4 (when poured), 3.1 to 4 (when blown)
Vermiculite loose-fill	2 to 2.6
Perlite loose-fill	2 to 2.7
Polystyrene rigid	4 to 5.4
Polyurethane rigid	6.7 to 8
Polyisocyanurate rigid	8

FIGURE 8.31 R-values for insulation.

Did You Know: that full inspections from code officers are needed for attic conversions? I would assume that you know this, but if you don't, know it now. Treat an attic conversion like new construction when it comes to inspections, and remember that plumbing, heating, and electrical systems must be inspected and approved by a code officer before you conceal the work.

Material	R-value per inch or as specified
Concrete	0.11
Mortar	0.20
Brick	0.20
Concrete block	1.11 for 8"
Softwood	1.25
Hardwood	0.090
Plywood	1.25
Hardboard	0.75
Glass	0.88 for single thickness
Double-pane insulated glass	1.72–⅜ w¼" air space
Air space (vertical)	1.35–¾"
Gypsum lath and plaster	0.40–⅞"
Dry wall	0.35–½"
Asphalt shingles	0.45
Wood shingles	0.95
Slate	0.05
Carpet	2.08
Vinyl flooring	0.05

FIGURE 8.32 R-values for common building materials.

FINISH WORK

The finish work needed for walls, ceilings, flooring, doors, and trim is all done in typical fashion. There are no special considerations for this type of work when remodeling an attic.

QUESTIONS AND ANSWERS

Q: I'm concerned about taking on an attic job. One of my major concerns is damaging the ceilings below the attic. Am I being too cautious?

A: *No. There is a high level of risk to the ceilings below an attic job. Be careful not to bang around too hard, follow the instructions in this chapter, and you should be fine.*

Q: Can't I just build over existing joists in an attic?

A: *You can't count on it. It's common for the joists in attics to be inadequate for living space. You have to confirm that the joists will support the load requirements of the use that you are converting the attic for.*

Q: Should I put dormers in for any attic I convert?

A: *Not necessarily. The use of the attic will dictate the need for windows in the roof line.*

Q: Should I remove existing insulation between the existing living space and the new attic floor?

A: *Taking the insulation out will help with mechanical installations and will allow more natural heat rise into the living space. Don't remove insulation that will be between downstairs living space and an unheated area behind the knee walls of an attic.*

Q: What is the best way for me to get 4 x 8 sheets of material into the attic?

A: *I favor using a gable window opening to bring the material into the attic. If there is not one in place, rough one in.*

Q: How do you feel about throwing debris down from an attic to a temporary trash bin?

A: *It can be risky. Someone could be hit by falling debris. I would prefer building a chute to contain the debris on the way down.*

Chapter 9

DORMER INSTALLATIONS

Dormer installations are common. They are built into new homes and added to older homes. The work is specialized. Building a dormer can be challenging, but the job is not one that should scare experienced contractors. Cutting into a roof and building new space out from the roof can be a bit daunting. Even so, the work is fairly routine and does not offer many major problems.

The type of dormer that you will be building will influence the amount of work at hand (Figure 9.1). A gable dormer requires far less work than a shed dormer. Small, gable dormers are not nearly as large as most shed dormers. However, there is a lot of detail work that goes into creating a gable dormer. This type of dormer is normally used to provide a place for a window.

Shed dormers are built to provide additional space in an

attic or upper living area. It is very common to build a shed dormer to house a bathroom. Building a full shed dormer, one that runs the entire width of a home, is not unusual. As you might imagine, this is a lot of work.

Dormers are frequently built during attic conversions (Figures 9.2, 9.3). They are needed for windows. When a shed dormer is installed, the square footage of living space is greatly increased. Whenever you open a roof, you have weather considerations to take into account. This is especially true when building a shed dormer, due to its size. So, let's start with weather protection.

WEATHER PROTECTION

Weather protection is a concern when adding a dormer to an existing home. The roof has to be cut open to install a dormer. This means that the interior of the home is at risk from rain, sleet, or snow. Openings for gable dormers are smaller, so the protection process is easier. Full dormers require more thought and preparation.

A general purpose tarp is all that is needed for weather protection in most cases. But, the tarp must be secured very well to prevent it from being blown up or away by high winds. When a roof is opened, it should be covered when crews are not working at the site. Assuming that a roof has an adequate pitch, a trap can be laid across the opening and secured with wood strips.

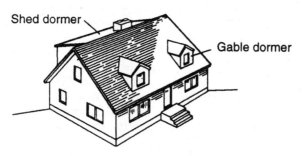

FIGURE 9.1 Illustration of gable dormers and shed dormer.

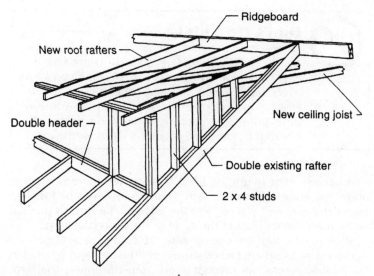

FIGURE 9.2 Framing for a shed dormer.

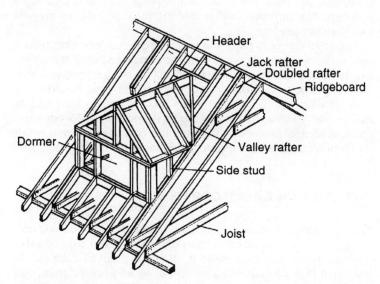

FIGURE 9.3 Framing for a gable dormer.

> **Don't Do This!** It can be very tempting to ignore sealing a roof opening when you are just going out for supplies or to lunch. Don't take this chance. Sudden downpours do occur. If you leave a job, secure the roof opening before you go. The extra work will often be what seems a waste of time, but the one occasion when it saves the interior of the home will make the effort worthwhile.

A second tarp might be used to keep water from running under the edge of the main tarp. Flashing can even be installed under shingles and placed over the edge of the tarp to prevent water from running under the edge of the protective tarp.

Wind is the biggest enemy when it comes to a tarp. Thin wooden strips that are tacked down over the tarp and secured in the roof structure will prevent wind from disrupting the tarp. When you are dealing with a small roof opening, like the one used for a gable dormer, this should be all that is needed. However, if the roof is fairly flat, you should tack in a piece of plywood under the tart. You don't want water pooling up and creating a problem. The firm plywood under the tarp will help to prevent this from happening.

Full shed dormers require a much larger roof opening than a gable dormer does. The increased size makes sealing the opening more of a challenge. A tarp will still do the job, but you may have to provide more backing for the tarp than you would if you were dealing with a gable dormer. Tacking boards to rafters or trusses will give you some backing to support a tarp. Aside from the overall size, protecting the opening of a shed dormer is not much different than closing up an opening for a gable dormer.

CUTTING IN A GABLE DORMER

Cutting in a gable dormer is not a big deal. In fact, you will normally be cutting out no more than one rafter. If you cut out a section of a rafter, you must head it off at the point of each cut. In the event that you cut out a section of an engineered truss, you should get an official ruling from an engineer about what is needed to provide adequate support for the compromised truss.

The general process when rafters are used is a simple header placed to butt into the cut edge of the existing rafter. The header may be doubled for added strength.

Since there is usually only one rafter cut for a gable dormer, there is not a lot of need for temporary support during the cutting phase. If you want to be extremely cautious, you can install temporary props under the ridgeboard to keep it from sagging when the rafter is cut (Figure 9.4). This doesn't take long or cost much, and it is added insurance of a job done well.

CUTTING IN A SHED DORMER

Cutting in a shed dormer can mean doing a lot of damage to an existing roof structure (Figure 9.5). Due to the size of shed dormers, many rafters or trusses must be cut. This will have a definite effect on the structural integrity of a roof structure. Therefore, you must take preliminary measures to support the roof structure before any structural cuts are done.

How will you support the roof? You can prop up the ridgeboard for starters. It is also wise to build temporary kneewalls to give added strength to rafters or trusses that will not be cut. In extreme situations, you should seek the expert advice of a structural engineer to ensure that you will not be creating a disaster when you open up a roof for a shed dormer.

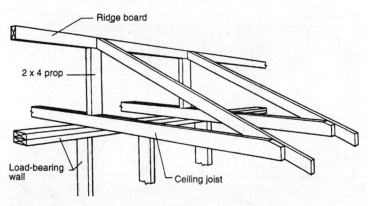

FIGURE 9.4 A temporary prop used to support a ridgeboard.

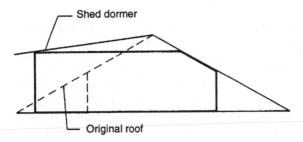

FIGURE 9.5 Profile of a shed dormer.

Shed dormers are built in various sizes. One of the most common types of shed dormers is built to house a bathroom (Figure 9.6). The space required for a full bathroom is not extensive. A room with dimensions of 5 x 7 feet will provide adequate space for a full bathroom. A shed dormer of this size is not much more trouble to deal with than a gable dormer is. In fact, plenty of carpenters would prefer to build this type of dormer. The detail work required for a gable dormer is more time consuming than the work necessary for framing a shed dormer of small dimensions.

If a full dormer is required, the work is extensive. These dormers can range in size from 24 to 40 feet, or more. This is a substantial undertaking. Cutting a major hole in one side of the roof of a house is nothing to be taken lightly. Make sure that the roof

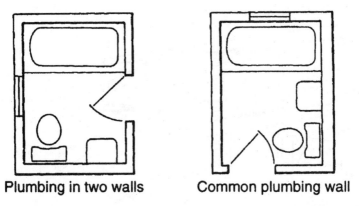

FIGURE 9.6 Typical bathroom layouts in a small space.

> **Don't Do This!** Don't allow roofing rip-out material to be piled up in a yard. People can be hurt by stepping on nails. Children may climb on the pile. Coordinate your cut-out work with a debris-removal plan. Either secure the debris in an on-site container, or haul it away.

structure is supported properly. Temporary supports and bracing will be needed to provide weather protection.

As with any roofing work, you have to create a safe way to get debris to the ground. The area below the work area should be sealed off to prevent injuries. This is true of any roofing work. Once the area is secure and the existing roof structure is supported for remodeling work, the cutting process can begin.

FRAMING

The framing of a dormer varies depending upon the type of dormer being installed (Figure 9.7). Even though shed dormers are larger than gable dormers, most contractors find the framing for a shed dormer to be easier and faster. There is a lot of detail work that goes into a gable dormer. When you consider the time spent on a gable dormer for the small amount of space gained, it does seem like a pain.

Shed dormers are often much larger then gable dormers (Figure 9.8). I've said this before, but it is a factor. It is not uncommon for a full shed dormer to span the width of a house. The type of framing required for this type of job is considerably different from the framing for a gable dormer. The exterior wall of the dormer will usually rest over the exterior wall for the lower portion of the home. Roof pitch is a factor, and it can be a problem.

Roof Pitch

Roof pitch is not a problem with a gable dormer, but it can be a problem with a shed dormer. The existing roof pitch can restrict the amount of pitch that you can put on a shed dormer. This can result in an addition that doesn't look good. It can also compromise the

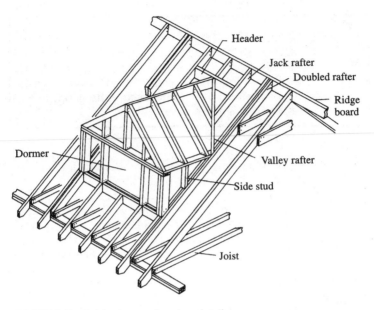

FIGURE 9.7 Gable dormer framing detail.

FIGURE 9.8 Full shed dormer.

load of the dormer roof. Additionally, you will have to give special consideration to the roofing material used if the pitch is not steep enough (Figures 9.9, 9.10).

If you are working in a region where cold climates are common, you have to take ice dams into consideration. The lower the roof pitch is, the higher the risk for an ice dam. Preventing ice dams is a combination of flashing and ventilation (Figure 9.11). It is common to install membrane flashing under shingles to prevent roof damage from ice dams. Metal flashing is also used. How much attention is needed to this problem? It depends on your regional location. The proper combination of flashing, ventilation, and insulation will prevent many potential problems.

Extending the length of rafter tails can provide both cosmetic and practical benefits. The extension is usually seen as a cosmetic

Traditional	Metric
2/12	50/300
4/12	100/300
6/12	150/300
8/12	200/300
10/12	250/300
12/12	300/300

FIGURE 9.9 Common roof pitches.

Material	Slope
Asphalt or fiberglass shingle	4 in 12 slope
Roll roofing with exposed nails	3 in 12 slope
Roll roofing with concealed nails 3" head lap	2 in 12 slope
Double coverage half lap	1 in 12 slope
Lower slope: Treat as a flat roof. Use a continuous membrane system: either built up felt/asphalt with crushed stone or metal system with sealed or soldered seams.	

FIGURE 9.10 Lowest permissible slopes for roofing materials.

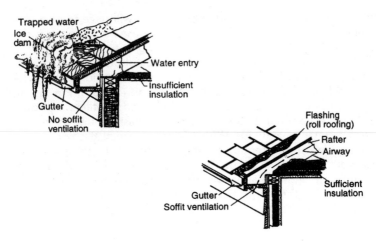

FIGURE 9.11 Details of how to prevent ice dams.

element (Figure 9.12). However, when a roof is extended past the exterior wall, it changes the drip line of water running off of the roof. This is more important with primary roofs than it is for dormer roofs.

WINDOWS

Windows are a common reason for building a gable dormer. This is basically the only way to get windows into the roof area of an attic area, unless roof windows are used. When shed dormers are used, windows are often installed. Regardless of whether you are building a gable dormer or a shed dormer, the window framing and installation is essentially the same as it would be with any new construction (Figures 9.13–9.15).

SIDING

Installing siding for a dormer is not much different than it would be for a typical wall (Figure 9.16). One exception on this point would be where siding meets the roof structure and shingles.

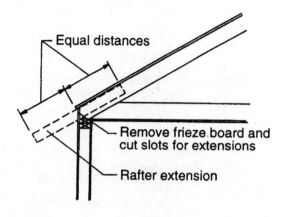

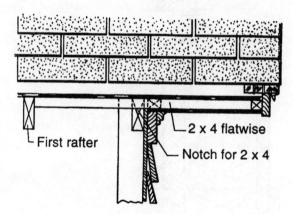

FIGURE 9.12 Details of roof extension.

The type of siding used for dormers can be the same types used for other structures (Figures 9.17–9.20). Beveled siding is very popular, but any type of siding is suitable for a dormer.

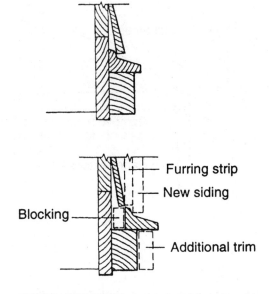

FIGURE 9.13 Changes in the drip cap with new siding.

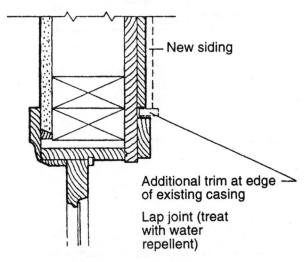

FIGURE 9.14 Top view of window casing extended for new siding by adding trim at the edge of existing casing.

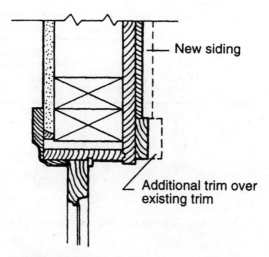

FIGURE 9.15 Top view of window casing extended by adding trim over existing trim.

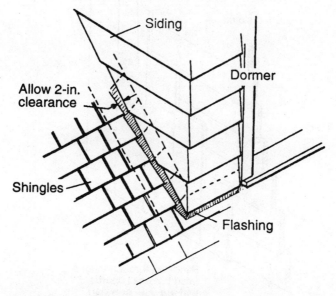

FIGURE 9.16 Flashing and siding clearance at dormer walls.

		Maximum exposure (in.)		
			Double coursing	
Material	Length of material (in.)	Single coursing	No. 1 grade	No. 2 grade
Shingles	16	7-1/2	12	10
	18	8-1/2	14	11
	24	11-1/2	16	14
Shakes	18	8-1/2	14	–
(handsplit	24	11-1/2	20	–
and resawn)	32	15	–	–

FIGURE 9.17 Exposure distances for wood shingles and hakes on side walls.

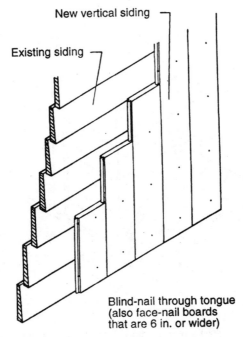

FIGURE 9.18 Details of applying vertical siding.

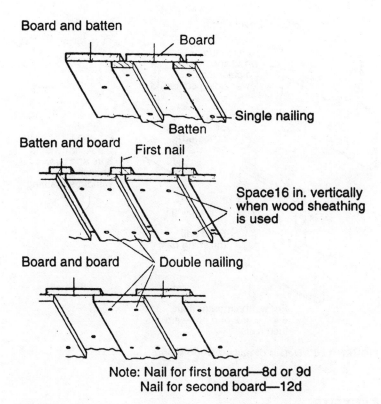

FIGURE 9.19 Details of applying board-and-batten siding.

TRIM

The trim work for a dormer is no different than it is for new construction. You have full flexibility in this area of dormer creation.

PAINT

Paint or stain can be applied to the siding for dormers. As with many elements of dormer construction, the painting or staining is the same as it would be for any other type of siding job.

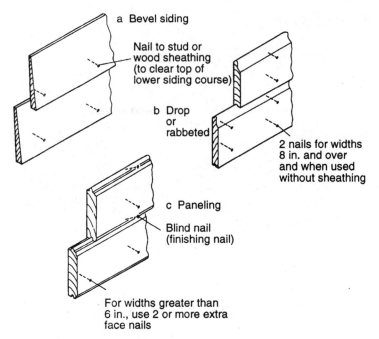

FIGURE 9.20 Details of nailing wood siding.

BENEFITS

The benefits of dormers are great. Whether you are adding room for a single window or creating space for bedrooms and bathrooms, dormers are the answer when it comes to attic conversions. The work involved with dormers can be frustrating and expensive, but the results are generally worth the effort. If you are involved in an attic conversion, don't overlook the use of dormers.

QUESTIONS AND ANSWERS

Q: What do most people think of dormers?

A: *I find that most people feel that dormers are fine. In fact, I can't remember a time when anyone complained about the appearance of a dormer.*

Q: Aren't dormers very expensive when they are used only for windows?

A: Yes, they are, but they are the only reasonable way to accommodate traditional windows in the roof of an attic.

Q: How important is the flashing for a gable dormer?

A: It's critical. Without the proper flashing, there will be leaks. This, of course, is not a situation that you will want to deal with once your job is done.

Q: Should laundry facilities be placed in attic conversions?

A: I don't like this type of setup, but it can be done. There is a risk of flooding lower floors if the hoses for a washing machine break. At a minimum, you should install a safety pan under the washing machine with a drain that will minimize water damage if a leak occurs. Personally, I think the safety plan will work for plumbing drips and oil leaks, but if a washer hose ruptures, there is going to be flooding. I try to avoid this type of laundry placement.

Q: Can I build a raised platform for a bathroom to make up for a lack of grade on drainpipes?

A: If you have adequate headspace, you can. However, the elevated platform can lead to tripping and falling accidents. I don't recommend this procedure.

Q: Should I have a pest inspection done on an attic before I remodel it?

A: This is both an excellent question and an outstanding idea. Powder-post beetles and woodborers can create terrible damage in an attic. Even termites can attack an attic. Normally, the damage is obvious to experienced remodelers, but you can't go wrong by ordering a professional pest inspection before investing time and money in an attic conversion.

Chapter 10

ENCLOSED PORCH CONVERSIONS

Enclosed porch conversions can be done to create many different types of use. The process can involve framing exterior walls and basically building the room structure under an existing porch roof. Or, you might have a job where an existing enclosed porch is remodeled to change its use.

Porches have been enclosed and converted for a multitude of uses. They are sometimes done to facilitate laundry facilities. A porch might be enclosed to provide a bedroom, a bathroom, a den, or a studio. The opportunities for potential use are extensive. A frequent reason for enclosing and changing the use of a porch is to establish a home office. Since porches can be accessed without going through a home, a porch makes good sense for a home-based worker who has clients or customers stopping by.

Instructions given in other chapters, such as Chapter 6, provide basic framing and foundation information. Siding, windows, doors, and trim for a porch conversion are dealt with as they would be with any new construction or remodeling job. There are, however, certain considerations to take into account once you know the nature of the conversion and use. This will be our primary focus in this chapter.

SIMPLE CONVERSIONS

Simple conversions can be done quickly. This type of conversion often requires nothing more than adding or removing some interior partitions, upgrading flooring, basic interior wall finishing, and painting.

When you have an existing enclosed porch to work with, you may not have a lot of work to do. Occasionally, you may have to provide additional electrical work, extended heating work, or maybe even plumbing. In most cases, the conversion will be little more than cosmetic work.

CLOSING IN A PORCH

Closing in a porch requires basic framing and remodeling skills. There are some things to check for before you give your customer a price for this type of work. Here are some examples of what you might want to look for:

Preliminary Considerations

- ✔ Is the foundation suitable for a new use?
- ✔ What is the condition of the existing roof?
- ✔ Will the slope of the existing roof permit the desired use?
- ✔ Can the existing floor structure be used?
- ✔ How difficult will it be to get heating, cooling, electrical, and plumbing into the new space?
- ✔ What will be involved in attaching the new walls to the existing home?
- ✔ Does the width or length of the existing porch need to be expanded?

Did You Know: that the conversion of an existing porch can give a homeowner substantial living space at a very affordable cost? Depending on the type of porch and conditions that you are working with, the enclosing of a porch can be very cost effective.

Assuming that you can use the existing roof, foundation, and floor structure, the framing work is not very complicated. You will have to check existing conditions to see that they are level and plumb. The odds are high that they will not be. This will require you to make adjustments in your framing. In some cases, you may have to use screeds to build a level floor surface. Once you have working conditions that will allow you to work effectively, the process of framing the exterior walls will be about the same as you would use for new construction.

It is very common for porches to have a sloped floor. This is something you should always check closely before beginning the framing work. If the porch you will be enclosing has a wood floor, check the structural condition of both the flooring and the joists. The last thing you want to do is frame up a porch and finish it off only to find out that the floor's structural system is inadequate.

Temporary Bracing

Temporary bracing may be needed when you are working in the framing stage. If the existing porch has large pillars that will not be hidden in the new framing, you must install temporary bracing to support the roof structure while you remove the pillars. One way to do this is with temporary posts. Another method is to put sections

Trade Tip: When using screeds on concrete flooring, be sure to use pressure-treated lumber. The moisture that lumber can pick up from concrete may cause the lumber to rot if you don't use treated material.

> **Don't Do This!** Some contractors come across a porch with a wood floor and are satisfied to probe the wood with an ice pick or screwdriver. Don't rely on this procedure. You must verify the size and spacing of joists and girders to ensure that you will not have code problems or a failure of the system.

of the outside wall together and stand them up between the pillars to support the roof as the pillars are removed. A third means of support can include using prop posts that are placed at an angle from the outside of the porch to support the roof during the framing process. Just don't make the mistake of removing existing support before having some form of suitable temporary support in place.

Insulation

You will normally have to add insulation to the attic of the porch roof. Depending on the type of roof you are working with, blowing loose fill insulation into the attic will probably make the most sense. Just remember that insulation wasn't needed when the roof was over a porch, but it will be needed as the transformation into living space is completed.

Steps

Not all porches or covered patios have suitable steps. Look for this early in your evaluation stage. Most porches will have some type of step system, but confirm that they will meet the new needs of the converted space. If you are going to pour a footing pad for the steps, it's a good idea to do it before you go beyond the rough framing. This reduces the likelihood of having a concrete chute knock out a window or damage new siding.

TYPE OF USE

Before you get too involved with the planning of a porch conversion, find out what the proposed use for the space is. This

Trade Tip: When you enclose a porch, expect to add some additional electrical wiring. This goes without saying in most cases, but by enclosing the porch and converting it to living space, you open the door to added electrical requirements, such as an exterior light that may not be in place when you begin.

information will help you to calculate what will be required to make the conversion successful. I have seen porches converted for a number of reasons. Here are a few types of uses that you may find a customer has in mind for a porch:

- Bedroom
- Bathroom
- Sewing room
- Den or study
- Laundry room
- Photography darkroom
- Miniature in-law addition
- Greenhouse
- Play room
- Music room
- Home office
- Woodworking shop

The list of potential uses for a converted porch can be a very long one. Each type of use can affect the remodeling requirements. To illustrate this, let's talk about some types of uses and see how the remodeling needs vary from use to use.

HOME OFFICES

Home offices are a very common use for converted porches. Offices have various needs. In today's computer age, wiring requirements are one of the most important elements needed for an office over some other types of uses. This includes both electrical and phone wiring.

Ideally, there should be several electrical circuits in an office. There should also be plenty of electrical outlets. Before you do a basic wiring job on a conversion that will be used as an office, sit down with the customers and draw a detailed electrical plan. Find out what they know they will need and then add some additional wiring for changes in the plan or growth.

Trade Tip: If you want to provide your customers with affordable natural light in a manner that will not compromise wall space, consider using fixed-glass panels. The panels are affordable and can be made to custom specifications. Putting a perimeter of glass panels around a room, near where the walls meet the ceiling, will give a modern, stylish look that offers a lot of natural light.

Plumbing

Plumbing requirements for an office may be minimal. If space allows, your customer may want a half-bath accessible in the office area. A wet bar might be requested. Otherwise, plumbing needs should be limited.

Lighting

Lighting can be a big issue in an office. I have found that track lighting is both practical and appreciated in small office environments. You might consider installing skylights to give added natural light.

Doors

The location and number of doors installed for a home office can be different than what would be experienced for a bedroom or a playroom. Traffic patterns are always important, so pay close attention to how people will move in, out, and about the room.

Flooring

The type of flooring used in an office may need to be more resistant to heavy traffic than what would be installed in a bedroom. If the office will have people coming and going often, a commercial grade of carpet may make the most sense as a flooring choice. On the other hand, if the office will simply provide a private place for a homeowner to work alone, the flooring choices and options expand.

Built-In Shelves

Built-in shelves for books or general storage is a good idea for an office conversion. This conserves floor space while providing storage options. Another advantage is that the built-in units provide a larger visual concept for the room.

Wall Surfaces

Wall surfaces in an office environment may suffer from moving furniture, rolling chairs, and other common office encounters. A strong, wood wainscoting on the lower portion of office walls can prevent dings in the walls. This type of decision is, of course, made by the customer, but it is the remodeler's job to aid customers in making good decisions.

Window Placement

Window placement should be given strong consideration when a porch conversion will be used as an office. Filing cabinets, high-back desks, bookcases, and other office furniture can interfere with window locations. In some cases, it's not a bad idea to use awning windows that will install above the furniture height. This probably will not be necessary, but you will have to do a mock layout of the office plan to determine where windows should be installed.

BEDROOMS

Porches are sometimes converted for use as bedrooms. This is a simple use that does not require much in the way of special needs. Sticking to typical protocol for bedrooms will usually work fine with a porch conversion.

BATHROOMS

New plumbing is needed when bathrooms are added to porch conversions. This can be a problem. You will have to confirm that the existing building drain and sewer is large enough to accept the new load of fixture units. A ground-fault circuit will be needed when wiring the bathroom. Venting for a bathroom fan may be needed. You will also have to assess the potential routes for installing new plumbing. It can be difficult to obtain enough grade on drains when connecting them to existing drains. Basically, you have to look at all of the elements that you would for the addition of any bathroom. It is not the porch itself that makes the plumbing potentially problematic, but rather the common problems that can block any bathroom addition.

DARKROOMS

Photography darkrooms require special ventilation, plenty of electrical outlets, and plumbing. All of this can create trouble for a remodeler who doesn't plan on the needs in advance. The plumbing is often nothing more than a sink, so the risk of overloading an existing plumbing system is low. You can even use an under-sink pump to pump the waste up hill if needed to get the drain to an existing drain. Ventilation isn't a problem, so long as you rough it in during the framing process. As with any electrical addition, you have to make sure that the existing service panel has room for the circuits needed to outfit the darkroom.

LAUNDRY ROOMS

Porches are sometimes converted into laundry rooms. This is sometimes done to remove the need to carry laundry up or down stairs. A laundry room will need plumbing, but this is not the type of plumbing that is likely to overload an existing plumbing system.

The electrical needs for a laundry room require a large circuit for the clothes dryer. As long as the existing service panel is large enough and has open slots, the installation of the wiring is usually not a problem.

Noise Reduction

Noise reduction may be needed on the common wall between the existing house and the laundry area. As most people know, washers and dryers are not quiet appliances. To avoid disturbances in the main part of a home, you can sound deaden the

Trade Tip: When installing plumbing in a porch conversion you should avoid placing the plumbing pipes in exterior walls if you are working in a region where freezing temperatures may freeze the plumbing.

common wall with insulation. This is not very expensive, but the benefits will be greatly appreciated for years to come once the laundry room is put into use.

There are many ways to reduce the transfer of noise. A common method is to build a 2 x 6 wall and stuff it full of glass fiber insulation. Foam insulation boards are sometimes used. I have even seen acoustical tile applied on the common wall.

DEFINING THE USE

Define the desired use of a porch conversion before you plan your job. A den, study, or music room will not require much more than a bedroom conversion. But, if you are dealing with a special use, such as a hobby room, you might need to provide additional lighting, ventilation, and electrical outlets. Noise reduction may also be wanted in a hobby room. People sitting in a family room watching television probably will not like it when a roar of a table saw shatters their living space. The more that you can take these types of matters into consideration, the better your odds are of winding up with happy customers.

QUESTIONS AND ANSWERS

Q: Should I insulate the floor of an enclosed porch?
A: *Yes.*

Q: Do porches really make viable living space?
A: *It depends on the porch, but they usually do.*

Q: How much money should a person invest in a porch conversion?

A: *There are far too many variables to answer this question. A remodeling job is not always based on cost versus financial value. The decision is up to the customer.*

Q: Do many people still use awning windows?

A: *No.*

Q: Does an enclosed porch count as living space on a residential real estate appraisal?

A: *If the porch is equipped with similar amenities to the home and attached access, it should count as living space.*

Chapter 11

GARAGES

Garages can be goldmines for remodeling contractors. Whether you are building a new garage, converting an existing garage, or adding living space over a garage, you can find a lot of work in this field.

Many first-time homebuyers cannot afford a garage when they buy their homes. This does not mean that they would not like one, only that the added cost of a garage makes the acquisition either uncomfortable or impossible. When these property owners save more money, they often opt to have a garage built. This is where the remodeling contractor comes in.

When I ran my company in Virginia, I maintained a garage crew. It was not uncommon during peak times for us to build one garage a week. My company also did a lot of conversion work. The most popular type of this work was creating in-law suites over garages. These apartment-type spaces were also popular with families who had teenagers who wanted a place of their own. I've been in Maine for the last 15 years, and garages are still a profitable side of the remodeling business.

The new construction of garages is appealing because the structure can be built quickly and the profit percentage is very good. Converting second-story living space that was framed for this intent is also fairly fast work. But, if you get into removing an existing roof, framing for living space, and finishing the space, the work will take much more time.

DETACHED OR ATTACHED?

If you are going to build a new garage, you and your customers will have to decide if the new garage will be detached or attached. There are several factors that can come into play when making this decision:

- Is there adequate room to construct a freestanding, detached garage?
- Will an attached garage be feasible?
- Can you get a suitable roof pitch with an attached garage?
- Will an attached garage be able to be fitted with a connecting door between the garage and the home so that the door opens into a suitable space?
- Is the customer willing to walk through the weather elements to get from a detached garage to the home?
- Are there underground utilities that will affect the placement of a garage?
- Is the customer willing to give up existing windows along the wall where a garage would be attached?

The questions in the checklist above are some samples of what should be considered before making a decision for a particular type of garage. People generally prefer attached garages, but this is not always the case. There are legitimate reasons for wanting a garage to be detached. Here are a few of them:

- A detached garage keeps gasoline away from the living space of a home.

- Garages that will be used for woodworking, or other types of work where loud noises will occur, might be built as detached units to minimize disturbance in the home.
- An attached garage that has living space above it can affect the view from both the lower and upper level of a home.

ZONING

We have talked about the need to verify how zoning laws, covenants, and restrictions may prevent you from building an outside structure. This applies to garages as much as it does to room additions. Since we have already covered this topic, I will simply state that you must confirm the approvals that you will need to build a garage or to alter the exterior appearance of an existing garage.

SOIL CONDITIONS

Soil conditions are a consideration for any foundation (Figure 11.1). If you are building a new garage, you will have to give consideration to the existing soil conditions. The odds are good

Material	Support capability
Hard rock	80 tons
Loose rock	20 tons
Hardpan	10 tons
Gravel	6 tons
Coarse dry sand	3 tons
Hard clay	4 tons
Fine dry sand	3 tons
Mixed sand and clay	2 tons
Wet sand	2 tons
Firm clay	2 tons
Soft clay	1 ton

FIGURE 11.1 Bearing capacities of foundation soils.

that the soil will be fine, but you should have it checked out. You certainly do not want a garage sinking into the soil after construction.

Finding the area of a footing can be done with a fairly simple calculation. The load divided by soil-bearing capacity will reveal the required number of square feet, or area, required of a foundation footing (Figures 11.2–11.4). For example, if you had a load of 60 tons per square foot, that would be built in soil with a capacity rating of 20, you would need 3 square feet of footing. This means that your footing would be 3 feet wide.

The drainage rating of a soil type can affect a garage. Since garages are built on slabs, you don't want a situation where hard rains will flood the garage. This may require the installation of slotted drainpipe to route water away from the garage foundation. If

Tons/sq. ft. of footing	Type of soil
1	Soft Clay
	Sandy loam
	Firm clay/sand
	Loose fine sand
2	Hard clay
	Compact fine sand
3	Sand/gravel
	Loose coarse sand
4	Compact coarse sand
	Loose gravel
6	Gravel
	Compact sand/gravel
8	Soft rock
10	Very compact
	Gravel and sand
15	Hard pan
	Hard shale
	Sandstone
25	Medium hard rock
40	Sound hard rock
100	Bedrock
	Granite
	Gneiss

FIGURE 11.2 Safe loads by soil type.

Type of soil	Percent of swell	Percent of shrinkage
Sand	14–16	12–14
Gravel	14–16	12–14
Loam	20	17
Common earth	25	20
Dense clay	33	25
Solid rock	50–75	0

FIGURE 11.3 Swelling and shrinkage of soils.

Soil type	Drainage rating	Frost heave potential	Expansion potential
Bedrock	Poor	Low	Low
Well-graded gravels	Good	Low	Low
Poorly graded gravels	Good	Low	Low
Well-graded sand	Good	Low	Low
Poorly graded sand	Good	Low	Low
Silty gravel	Good	Moderate	Low
Silty sand	Good	Moderate	Low
Clayey gravels	Moderate	Moderate	Low
Clayey sands	Moderate	Moderate	Low

FIGURE 11.4 Soil properties.

you suspect that the soil where the garage will be built will not drain well, take precautions to protect against flooding.

FOOTINGS

We have already talked about the construction of footings, so there is not much need to go over it again. However, there is one wrinkle that is worth talking about. Footings for rooms are usually intended for use with a wood flooring system. Since the floor for a garage is almost always made of concrete, this allows a

Did You Know: That the formula for estimating concrete is not very complicated? To determine the amount of concrete needed, you multiply the length by the width by the thickness. For example, a project that is 50 feet long, 10 feet wide, and 8 inches deep will require 333.33 cubic feet of concrete. Because concrete is sold in cubic yards, convert your findings by dividing the total (333.33) by 27. You will arrive at an answer of 12.35 cubic yards (Figure 11.5).

somewhat different approach. I am talking about a monolithic pour. This is when the slab and the footings are poured all at once and are integral units. A floating slab can be used, but many contractors favor a monolithic pour.

When making a monolithic pour, the footings and the slab area are both prepared for concrete before any concrete is poured. This means having the crushed stone and wire in place over the pad location before pouring the footings. Don't forget to have the needed inspections from the local code enforcement office prior to pouring the concrete.

The thickness of a garage slab varies (Figure 11.6). At a minimum, they are usually 4 inches thick. Some contractors prefer slabs that are 6 inches thick. The slab may have a thickness of 5 inches. The decision will depend on the use of the floor. If heavy equipment is going to be stored in a garage, a thicker floor is

Slab thickness (inches)	Slab area (square feet)				
	10	50	100	300	500
2	0.1	0.3	0.6	1.9	3.1
3	0.1	0.5	0.9	2.8	4.7
4	0.1	0.6	1.2	3.7	6.2
5	0.2	0.7	1.5	4.7	7.2
6	0.2	0.9	1.9	5.6	9.3

FIGURE 11.5 Estimating cubic yards of concrete.

Trade Tip: When you excavate for a garage footing and slab, go deep enough with the excavation to ensure that the foundation is on solid, undisturbed ground. I've seen jobs where contractors got fast and sloppy with the slab construction, and this can be a serious problem.

needed. However, most residential garages will do fine with a 4 inch slab. Check your local code requirements prior to making a final decision.

FLOOR FINISH

The floor finish for a garage may be a significant issue. Will your customer prefer a glassy finish or a swept finish on the concrete? Some customers will not like the slick finish that is most common in garages? Why? Because a smooth finish can become very slippery when it is wet. Some customers will prefer a swirled or swept finish that will allow good footing, even if the floor is wet.

GENERAL FRAMING

The general framing for a garage is essentially the same as it would be for a room addition or house. There are usually fewer windows in garages than what would be found in a house. The garage door

Use	Thickness
Basement floors in residences	4 inches
Garage floors for residential use	4 to 5 inches
Porch floors	4 to 5 inches
Base for tile flooring	2½ inches
Driveways	6 to 8 inches
Sidewalks	4 to 6 inches

FIGURE 11.6 Thickness of a concrete slab.

is an obvious difference when garage framing is compared to a room addition (Figure 11.7).

Most garages will have a garage door and a standard entry door. Some customers will require higher clearance than others. Take this into consideration before you frame the exterior walls. If your customer owns, or is considering owning a large vehicle, you might need to use longer wall studs to create the necessary clearance at the garage door (Figure 11.8).

Second story framing can be nothing more than typical attic framing (Figure 11.9). Some customers may want attic-storage

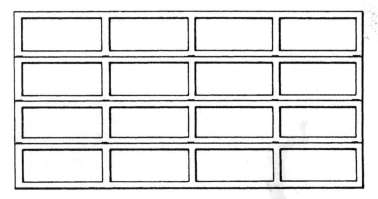

FIGURE 11.7 Sectional overhead garage door.

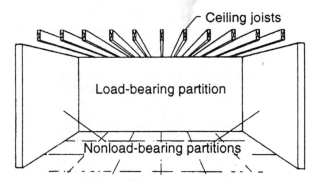

FIGURE 11.8 Diagram of both load-bearing and non-bearing partitions.

Garages

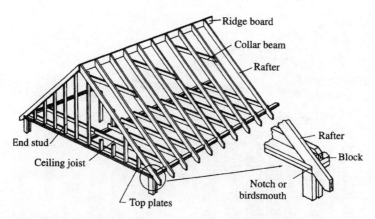

FIGURE 11.9 Standard attic framing for a gable roof.

trusses. Other customers will want living space above the garage. You can use rafters or trusses, depending upon the needs of the second-story space.

If the second story of a garage will require extensive strength to meet the needs of the use, you may have to install some steel beams or fitch plates to beef up the support (Figure 11.10). This will rarely be needed in a residential garage, but it is something to keep in mind.

When you are ready to frame the roof structure, you have to decide between trusses and rafters. Many contractors prefer trusses, due to the speed with which a roof can be put together when trusses are used (Figure 11.11). The trusses can be simple roof trusses, or they can be either attic-storage trusses or even room trusses. Using trusses does make the job go faster, but there are some limitations with trusses that you don't have when you stick build a roof.

Rafters allow you to build what you want, the way you want it, where you want it. This is one of the main reasons some contractors prefer the use of rafters (Figure 11.12). I have used both rafters and trusses and find that both of them have their place. The decision of which to use is largely a personal one, with site conditions sometimes dictating which will be best.

Sheathing a garage is done with the same procedures used on other roofs (Figure 11.13). Plywood or waferboard is normally used as sheathing. Remember that plywood clips are needed if the sheathing is not of a tongue-and-groove type.

FIGURE 11.10 The use of steel to strengthen the floor structure for a second story floor.

FIGURE 11.11 Standard roof trusses.

FIGURE 11.12 Standard rafters that have been cut and are ready for installation.

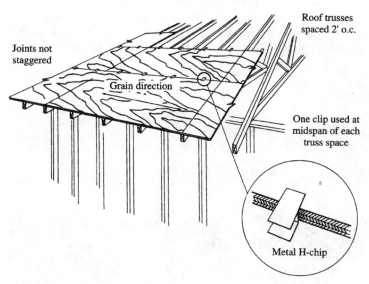

FIGURE 11.13 Sheathing a roof.

The shingles used on a garage roof are usually asphalt shingles, but any type of shingle can be used (Figure 11.14). Again, this phase of the work is so similar to any other type of roofing that it is not worth talking about again.

Interior framing is usually minimal in a garage. When partition walls are required, they are framed with the same methods used in other types of interior framing. Normally, the ground level of a garage does not include any interior framing. There may be a need for stairs to gain access to upstairs floor space. When this is the case, you have to lay out the stairs to avoid any conflict with the parking area inside of the garage. This simply requires basic planning on your part (Figures 11.15, 11.16).

MECHANICAL SYSTEMS

The mechanical systems in a garage can include plumbing, heating, cooling, and electrical systems. All garages will have electrical wiring (Figure 11.17). Some garages will have plumbing for hose bibs or floor drains. Heating and cooling systems are not common in garages, but there are many heated garages in regions where extremely cold climates exist.

In general, you can apply the same rules for mechanical systems to garages as you would use with a room addition. If the garage is attached to a home, the routing of mechanical systems

FIGURE 11.14 Shingles being applied to a garage roof.

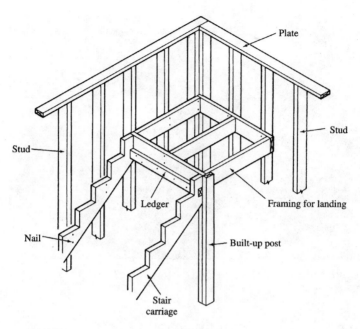

FIGURE 11.15 Typical stair framing.

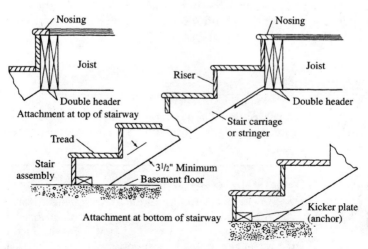

FIGURE 11.16 Methods of attaching the base of a set of stairs.

FIGURE 11.17 Typical wiring for a garage is no different than any other type of wiring.

can be considerably easier than if you are building a detached garage. However, a trench can provide a path for plumbing and electrical wiring to a detached garage. In some cases, it makes sense to set a garage up with its own electrical service and heating and cooling system. This, of course, will depend on the proposed use of the garage and the level of difficulty of tying into existing systems for services.

INSULATION

Insulation is usually optional in a garage. I suggest that you insulate the structure. As you know, insulation is not a major expense,

 Did You Know: that a ground-fault-interceptor circuit is usually required when electrical wiring is provided to a garage? Check your local codes for details, but I suspect that you will find that a GFI circuit will be needed.

but it does provide good service and is much easier to install during construction than it is after walls are covered.

WALL COVERING

The wall covering used in garages is normally drywall. Fire-resistant drywall is usually required on any common walls or ceilings between a garage and living space. The need for this fire protection may restrict some of your interior wall covering options, so make sure that you are within code compliance with your choice for a wall covering.

There are many garages that have exterior walls that are not finished on the interior. Bare studs are left as the final job. This is fine if it suits your customer. However, remember that common walls and ceilings between the garage and living space will need to be covered with an approved fire-rated wall covering.

EXTERIOR FINISHES

The exterior finishes for a garage can be about the same as you would use for a house or addition. For example, any type of siding and exterior trim can be used. Paint or stain can be applied to the siding. In general, you can treat the outside work of a garage as you would any other residential structure.

CONVERTING AN EXISTING GARAGE TO LIVING SPACE

When homeowners need more living space and have a limited budget, they may opt to have a garage converted to living space. This type of work is not particularly difficult. Since garage floors are typically lower than the floor level of a home, there may be a desire to build up the floor (Figure 11.18). This is not a bad idea. It gives you some space under the new floor to route mechanical systems and allows passage from the garage to the home without steps. You can accomplish this by simply building a new floor structure over the existing concrete slab.

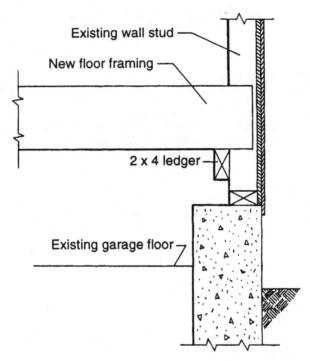

FIGURE 11.18 Framing to bring a floor in a garage conversion up to the same level as the floor level in a home.

If having a step, or two, into the garage is not a problem, you can use the existing floor. Or, you can use screeds to build the floor up for a wood floor. It is not unusual for garage floors to be build with a slight grade in them. Sometimes this grading goes to a floor drain. At other times the slant goes toward the garage door. This is a good reason to build up a false floor. And as I said earlier, the raised floor gives you room to work in your mechanical systems.

FRAMING IN THE GARAGE DOOR

Framing in the garage door is a logical step when a garage is converted to living space. All in all, this is not a big job. However, matching up the siding that will take the place of the door with the siding that is existing can be troublesome. From the framing point of view, you simply remove the door and frame up the opening. Once the framing is in place, you have to deal with matching the siding and the finished color. This can be tough.

Paint or stain that has aged can be very difficult to match. To avoid this, some contractors get creative. For example, you might install a sliding glass door and a couple of fixed-glass panels to replace the garage door. This will eliminate the need for new siding. You can fill the void with glass and trim it without making the change appear awkward.

You have to evaluate each job before deciding how to get around matching siding and colors. For another example, you might install a bank of windows and use some decorative type of siding to give the section a reason to look different. For example, you might create diamond shapes with new siding and seal it to maintain its natural color. This will stand out, but it will not look out of place.

MECHANICAL SYSTEMS

Mechanical systems installed in a garage conversion can be compared to those installed in a room addition. Since this topic was covered earlier, we will look past it here.

INSULATION

Not all garages are insulated. If the garage is to be converted to living space, insulation will be needed. Keep in mind that you will need the same types of inspections from the local code officers for garage conversions as you would for other types of construction.

WALL COVERINGS AND CEILINGS

Wall coverings and ceilings in a garage can be the same as those used for any other type of remodeling. All of the options are on the table.

FLOOR COVERINGS

Floor coverings for a garage conversion may be limited. If you are working directly on a concrete floor, you have numerous options. Even hardwood flooring can be used if you prep the floor properly. Since most contractors construct a wood floor over the concrete slab, there are no limitations.

In general, converting the ground level of a garage to living space is not a massive job. The complexity comes when you convert the space over the garage, or add space over a garage, to create living space.

GARAGE ATTICS

Converting garage attics to living space is not much different from doing a regular attic conversion. You can refer back to Chapter 8 for details on attic conversions. Garages are often framed in a way consistent with a routine attic. It's not unusual to find garages that have attic storage trusses. When this is the case, don't think that you can do some alterations to the trusses and call them good for rooms. If you alter any portion of a truss you run the risk of compromising its structural integrity. Even if you come across an attic with room trusses in it, you must not alter the trusses. Take the time and go to the expense of doing the job right.

Whether you are building new garages or converting existing garages, you are in a potentially lucrative business. Of course, most remodeling does offer the opportunity to earn a good living. The key is making and maintaining a good reputation. Don't take this part of your job lightly.

 Don't Do This! Some contractors are tempted to put finished living space over a garage without obtaining a permit to do the work. Occasionally, contractors will try to cheat the system to get away with substandard joists for the second floor. Don't do it. The risk is not worth the reward.

QUESTIONS AND ANSWERS

Q: Are sump pumps needed in garages?
A: *If you provide gutters and adequate drainage on the perimeter of a garage, a sump pump shouldn't be needed. This is not to say that you may never need a sump pump in a garage.*

Q: Can I just frame up a pad and pour a slab on grade for a garage?
A: *This is not a good idea. The slab will float with frost heaves and settle. A garage should be built on a suitable foundation.*

Q: Can I use steps built on the outside of a garage for access to the upstairs living area over the garage?
A: *Yes. However, exterior steps can be an eyesore and they can present safety hazards during icy weather.*

Q: Do most people expect wiring for automatic garage door openers?
A: *I believe that most people do want the option for an electronic garage door opener.*

Q: What do you think about using a breezeway between a detached garage and a home?
A: *This is a viable way to provide sheltered access from a detached garage to a home.*

Q: What is the best type of independent heating system to use in a detached garage?
A: *A heat pump is one viable option, and it is probably the best choice if air conditioning is wanted. An oil-fired heater that vents through the wall is a good way to provide heat if you don't need air conditioning.*

Q: Is it safe to sleep over a garage that houses a vehicle that has gasoline in its fuel tank?
A: *It would not be my favorite place to sleep, but with the proper precautions, it is legal. There are many homes that are built over garages when a building lot is heavily slopped and allows such construction.*

Chapter 12

DECKS, GAZEBOS, AND SCREENED PORCHES

Decks, screened porches, and gazebos are very common among residential communities. These home improvements offer a remodeling contractor an attractive means of making more money. Of the three, decks are the most popular. Many companies specialize in the construction of decks. In areas where population is dense, it is not unusual to find companies that do nothing but deck constructions.

The nice thing about building decks is that with the exception of attaching them to a home, you have a fresh start to work with. The work is not as frustrating as smoothing out an old plaster wall or fishing wire through old walls. Building decks is basically new construction, but the work can fall into a remodeler's realm when a deck is added to a home after the construction of the home has been completed. The same goes for gazebos and screened porches. Even if the porch or gazebo is not attached to a home, it is not uncommon for homeowners to call remodeling contractors to do the work.

Many contractors enjoy building decks. One reason for this is that the time required to complete deck construction is minimal when compared to other types of jobs. This allows a contractor to finish a job quickly, to get paid, and to move onto a new job. A lot of remodelers prefer fast jobs, and deck construction is quick work.

Gazebos come in very handy for outdoor living when biting insects are prolific. There are times when gazebos are attached to homes, but they are often freestanding. An octagonal design is typical for gazebos. Getting the right roof cuts for the rafters is one of the most challenging aspects of gazebo construction. Some gazebos are very plain and may be little more than a screen room. Other gazebos are very fancy and resemble a sunroom. There are a multitude of options possible when designing and planning the construction of a gazebo.

Screened porches are sometimes independent structures, but they are often attached to homes. Most screened porches have a deck floor, screen walls, and a routine roof. But, a screened porch can be considerably more than an enclosed deck. Like gazebos, screened porches can provide numerous options.

ZONING

Zoning, covenants, and restrictions can all come into play when building a deck, gazebo, or porch. We've talked about this issue earlier, but it is worth revisiting briefly. Whenever you add something to the exterior of a home or build a freestanding building, you must confirm that the construction will not violate zoning laws or recorded covenants and restrictions. Failure to do this can result in financial disaster.

Don't Do This! You might think that adding a small deck to a home is no big deal and that it will take too much time to research the legal requirements for exterior construction. Don't fall into this trap. Take the time to make sure that your work will not be in conflict with legal restrictions.

The rule of thumb for checking zoning laws and covenants and restrictions is that if you are altering the exterior of a building in any way or adding to the structure, or building a new structure, you should confirm any legal requirements pertaining to the proposed work.

Underground Obstacles

You may have to deal with underground obstacles when building a deck, porch, or gazebo. New structures usually have foundations. However, gazebos are sometimes built without a typical foundation.

If you are going to excavate ground for a foundation, you must make sure that there are no hidden obstructions in the area. A visual inspection is not enough. You might pick up on something obvious, such as a cleanout plug that indicates a sewer is in the area. But, a visual inspection will not always indicate buried piping or wiring.

Many communities have one number that contractors can call to gather details on buried utilities. If this service is available in your area, take advantage of it. Most phone directories list such numbers. If you are not sure of what is available, call your local code enforcement office and ask if there is a number that you can call for information on buried utilities.

THE FOUNDATION

The foundation of any structure is always very important (Figure 12.1). Decks are usually built on pier foundations. Screened porches and gazebos are also built on pier foundations from time to time. Sometimes, small gazebos are built on skids that allow the gazebo to be moved from one location to another. Trench footings are also used at times for porches and gazebos. All in all, I'd say that pier foundations are the most common when dealing with decks, screened porches, and gazebos.

If you use a pier foundation, the number of pier footings needed will depend on the size and design of the structure that you are building. Most decks attach to the homes that they serve. The attached section is usually attached to the home's band board with lag bolts. Piers are needed to support girders

Don't Do This! Do not rely too heavily on site plans that show underground utility locations. It is not uncommon on residential jobs for deviations to occur that are not recorded. For example, I've seen site plans that showed a septic tank location that was, in fact, on the opposite side of the lawn.

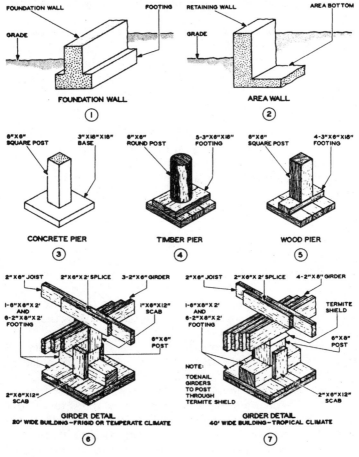

FIGURE 12.1 Types of foundations.

and the outside edges of the deck. This is often the same situation encountered with screened porches and gazebos.

Freestanding structures require more piers. The design for the foundation should be shown on the approved building plans. Permits are required from the code enforcement office for new structures. Typically, the code enforcement office will want detailed plans and specifications filed with an application for a building permit. Follow the requirements on the approved plans when creating the foundation for your structure.

Footings must be installed below the local frostline. You can check your local code book to determine what the minimum depth of a footing must be. It is also essential that footings are poured on solid, undisturbed ground. In some cases, this can result in much more digging that you might have planned for.

When I was a contractor in Virginia, the minimum depth of a footing was 18 inches. I remember one job where we had to dig down to more than 4 feet to find solid ground to pour concrete on. This type of a problem can add a lot of cost to a job. Not only do you have the additional excavation work to pay for, you will have increased cost in the foundation walls. This is not a big deal with a pier foundation, but it can eat at your profits if the foundation is a full wall.

Keep in mind that footings generally must be inspected by the local building inspector before they are filled with concrete. Once the footings are approved, you are able to pour the concrete and stock the job for framing.

ATTACHING TO AN EXISTING HOME

Attaching your framing to an existing home can be trickier than you might think. The concept is simple enough, but the execution can

Don't Do This! I have known contractors who tried to save money and time by building decks without the required building permits. This is bad business. Don't do it. Regardless of your thoughts about permits and inspections, the process protects you. Comply with your local building requirements at all times.

> **Trade Tip:** It is not a bad idea to limit your job bid to specific conditions. For example, you might include a clause in your bid that specifies the depth of the foundation and that if the soil conditions require a deeper foundation that the additional cost will be passed onto the customer as an additional charge.

be a bit more complex. You must first find your attachment location. This will usually be the band board of the existing structure. Once this is found, you have to cut away existing siding and get down to a flat surface. Flashing should be installed at the point of attachment. This will keep water from invading the attachment area. If water leaks behind the joining point, rot will ensue.

The wood member attached to the existing house is usually put in place with lag bolts. Some contractors use nails, but this is not recommended. Lag bolts that screw into the attachment area are far less likely to work loose than nails (Figure 12.2).

Once the attachment is made, you can establish the height of the new flooring area. This will tell you how tall the foundation walls or pier members will need to be. To get an accurate measure, you can use a string and a line level to reach any points where measurements are needed. I suggest driving a stake into the ground so that the string can be tied to the stake and secured at the bottom of the attached member. Stretch the line until it is tight. Confirm that the line level is, in fact, level. Then you can measure from the string to the footing to determine the height needed for proper framing support.

JOIST STRUCTURE

Once the footing and foundation work is done, you can move onto the joist structure (Figure 12.3). The methods used to begin framing the joist structure varies with the type of foundation and piers used. Some foundations are poured with pockets built into them for girders. Other styles use girders that sit on top of the supports.

There are several ways in which to make connections between sills and joists (Figure 12.4). The approved building plans should

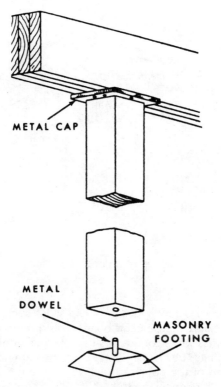

FIGURE 12.2 Diagram of a pier support used on a pier foundation.

show details of how you are authorized to make your connections. Even if you have a different means of connection in mind, remember that you must follow the approved plans to ensure a satisfactory building inspection.

Once the bands and girders are in place, you are ready to install the joists. Again, the size and spacing of the joists should be shown on the approved building plans (Figures 12.5–12.8). You can refer to tables to check allowable spans, but in any case, you should comply with the approved building plans and specifications.

Joists will rest on beams and be attached to the band boards. Some contractors attach the joists with only nails. I prefer to use joist hangers (Figure 12.9). You undoubtedly know what a joist hanger is, but in case you don't, they are simple metal devices that

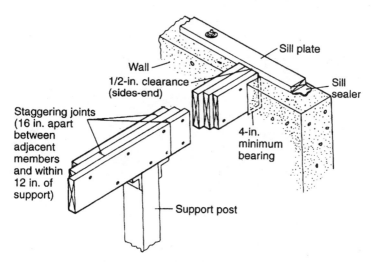

FIGURE 12.3 Typical built-up wood beam installation.

are installed from below a joist. The clip slides over the joist and the hanger is nailed to the band board. This gives additional support to the joists. Some contractors use a ledger strip to allow the joists to rest on it. The choice depends on the framing conditions.

The framing of the floor can take various forms, but the end result will be a strong structure that will accept and secure a flooring system (Figures 12.10, 12.11). Decking the flooring system is the next step in the construction process.

DECKING

The decking on a floor structure is not difficult. You and your customers have to decide what types of decking to use. I prefer a round-nosed, pressure treated decking for decks and screened enclosures. Some people prefer typical boards. Painting pressure treated lumber can be tough, and staining the wood is pretty much out of the question.

The dimensions of the decking can vary considerably. I've seen decks where the flooring was made with kiln-dried, plain 2 x 4s. I prefer a 5/4 round-nosed decking board. The choice is up to the customer. In any case, the installation is fairly simple.

Decks, Gazebos, and Screened Porches

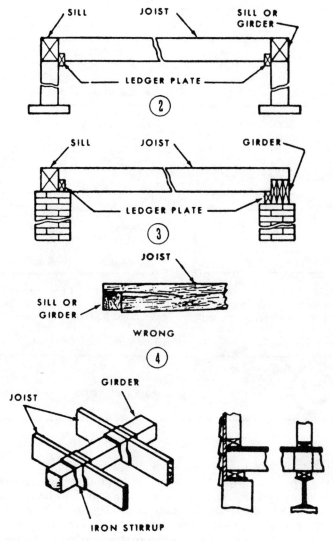

FIGURE 12.4 Sill and joist connections.

Beam size	Maximum distance between support post
4-x-6	6 feet
4-x-8	8 feet
4-x-10	10 feet
4-x-12	12 feet

FIGURE 12.5 Common beam spans for decks.

Beam size	Maximum span
2-x-6	8 feet
2-x-8	10 feet
2-x-10	13 feet

FIGURE 12.6 Common joist spans for decks where joists are installed 16 inches on center.

Beam size	Maximum span
2-x-6	6 feet
2-x-8	8 feet
2-x-10	10 feet

FIGURE 12.7 Common joist spans for decks where joists are installed 24 inches on center.

Beam size	Maximum span
2-x-6	5 feet
2-x-8	7 feet
2-x-10	8 feet

FIGURE 12.8 Common joist spans for decks where joists are installed 32 inches on center.

 Trade Tip: Use galvanized nails when installing decking to avoid rust marks.

It is important to establish a consistent spacing between the decking. A lot of carpenters use a nail to establish the spacing. They simply install one piece of decking and then hold a nail between it and the next piece of decking. This is a fast, effective method of maintaining suitable spacing.

When flooring a screened enclosure, the spacing in the decking is a problem. Many contractors overcome this problem by installing screen on the floor joists below the decking. The key is to keep insects out. Spacing is needed between the decking to prevent water from building up on the floor, so screening the flooring structure makes a lot of sense to let water out and to prevent pests from entering.

Decks, Gazebos, and Screened Porches 301

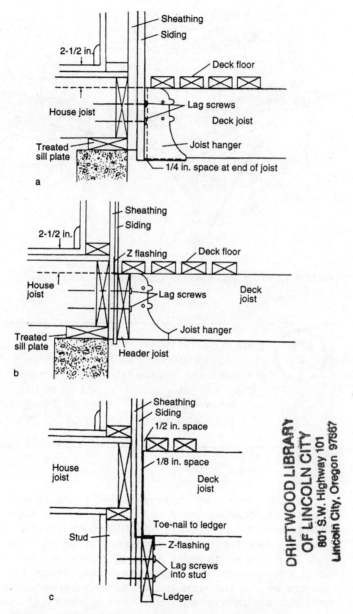

FIGURE 12.9 Common methods for attaching joist hangers and using ledger strips.

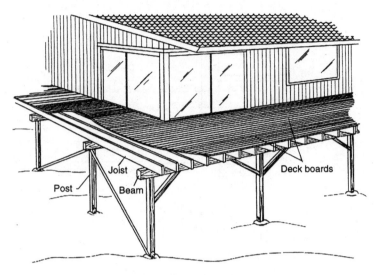

FIGURE 12.10 A typical pier foundation and flooring system.

 Did You Know: that customers will often pay extra for attractive patterns in decking if they are given the opportunity? My experience has shown that many customers will gladly pay more for a creative design in the flooring of their decks. Give this some thought before giving your next quote.

How often do you see decking installed in a straight pattern? Probably pretty often. But does it look good? Is it interesting? Will guests remember it? Well, straight decking is boring. Switch gears and install the decking diagonally. This will get some attention. If you want to build a strong reputation, incorporate designs. For example, you might have a center design that is surrounded by diagonal decking. Consider bringing decking diagonally from the four corners to the center to give a starburst look. If you perfect some special styles, your work will be talked about and more work should roll in.

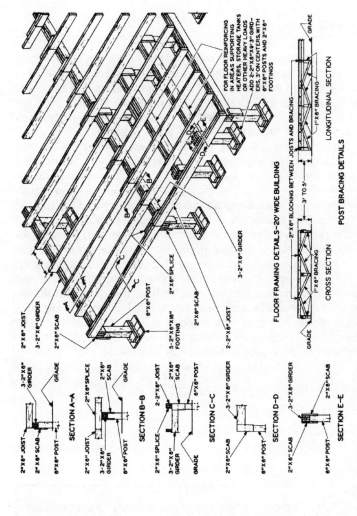

FIGURE 12.11 Floor framing details

Once the decking is done, you are ready to move into the framing of either rails or exterior walls, depending upon what you are building.

RAILING

If you are building a deck, the railing can take on many looks. When there is a tight budget required by the customer, you can use 2 x 4 stock for the top and bottom rail and simple pickets to meet code. Most customers will want what I call a drink rail. This is a 2 x 6 rail that accommodates glasses filled with beverages much better than a narrow rail will. The choice is up to your customer.

A typical railing will have a horizontal member at the bottom section and at the top section. Vertical pickets will be installed between the two horizontal rails to meet code requirements. Occasionally, contractors will forego the bottom rail and attach picked to the exterior band boards at the bottom and the top rail at the top. This is fine. It's a matter of appearance.

A simple railing will have a top rail, a bottom rail, and a number of square pickets that are nailed into place between the two rails. This is fine, but it is not flashy. Customers often want to make a statement with their decks. There is almost no end to the countless designs you can come up with for the code compliance required between the deck rails. My crews have built protective pickets in ways to create octagonal shapes, diamond designs, and so forth. Once again, having your own "signature" on your work will get your jobs talked about and create more work for you. It's some of the best advertising you can have.

Benches and Planters

Benches and flower planters are frequently built as an integral part of a deck and the railing system. Not everyone wants this

Trade Tip: Many contractors use untreated wood for rail posts, rails, and pickets. This allows more opportunity for painting or staining the wood components than treated wood allows.

type of amenity, but many people do. It's simple enough to incorporate either of these enhancements into the overall design of a deck and its railing system.

STEPS

Steps are usually needed for a deck or porch (Figure 12.12). The types of steps used can take on many facets, but pressure treated lumber is one of the mainstays in deck and porch steps. Some contractors don't provide footings for such steps. Instead, they rest the bottom of the stair risers on patio blocks. I prefer a true foundation for the step structure to rest on, but patio blocks are much less expensive and often work out very well.

The riser treads and railings for deck or porch steps do not have to be fancy. The use of pre-cut risers is often possible, and this saves a little time on the job. How much effort you put into making the work pretty is up to you and your customer.

DECKING FINISHES

The range of decking finishes is substantial. Many decks are built with pressure treated lumber and are left in that natural state. Some decks are floored with untreated wood and painted or stained. High-end decks use expensive wood decking and are protected with a clear protective coating. It is essential that untreated wood be protected with some type of protective covering. Personally, I don't like paint much for this type of work.

I normally use treated lumber for my jobs. The benefits are strong, and I see no reason to use untreated wood. However, there are customers who don't like the green look of treated wood. When this is the case, you have to go to other options. The bottom line usually comes down to what the customer wants. What you need to remember is to avoid leaving untreated wood in its natural state. If a customer insists on you doing this, get a disclaimer against rot and discoloration signed before you leave the job.

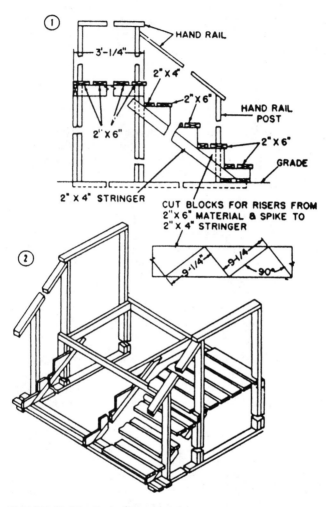

FIGURE 12.12 Typical step detail.

 Did You Know: that gazebos are very popular with people who have outdoor hot tubs? I've found that people who enjoy having their hot tubs outside of their homes love gazebos. Keep this in mind and you might extend your sales.

GAZEBOS

Gazebos offer more challenges than decks do. They are nothing to be intimidated by, but they do require some additional planning and skill to build. Many kits are available for gazebos. If you don't want to use the kits, you might like to take advantage of the pre-drawn blueprints that are available for gazebos. It is helpful to have detailed cut instructions.

A friend of mine is an excellent carpenter with years of experience. This past summer he built a gazebo for himself. The building is rectangular, but he put the type of roof on it that you would expect on an octagonal gazebo. During the summer, I spent some time on the job with him and was surprised that he was concerned about getting the rafter cuts right. This carpenter can build just about anything, but he had never done the type of roof that he wanted for his personal gazebo. As good as he is, it took him awhile to master the angles of the rafters. This is a good example of how experienced carpenters can benefit from some pre-designed kits or plans.

Once you get the formula for the rafters, the roofing system is not complicated. I strongly recommend the use of pre-planned blueprints. The entire concept of building an octagonal building scares away some carpenters. Building your first gazebo might be a bit of an undertaking, but once you have one under your belt, the rest of the ones you build will be much easier.

The uses I have seen for gazebos range from hot tubs to picnic areas. Many gazebos have only screen walls. Some buildings are framed so that glass panels can be inserted for winter use. This extends the seasonal use of the structure, and it is not difficult to frame the wall panels for this type of convertible use.

WALLS

The walls of a gazebo can be as simple as a screen covering. At the other end of the spectrum, a gazebo can have high-end herring-bone interior wall finish about half way up and then be enclosed with top-of-the-line casement windows. Cost is a factor in determining which route to take.

The basic framing of walls for a gazebo is not particularly difficult. There is a matter of mitering corners if you are doing an octagonal job. You will have to provide framed openings for windows if they are to be installed. If your intent is to simply screen the structure, you will only have to frame an opening for a door. Lattice strips will be used to conceal the joints where screen sections meet.

ELECTRICAL CONSIDERATIONS

What are the electrical considerations for a gazebo? Many gazebos have no electrical facilities. A gazebo where a hot tub will be installed will need a ground-fault-interceptor outlet or circuit. It is often desirable to have electricity in a gazebo. This can be for outlets, lights, and even a ceiling fan. If you anticipate significant wiring, plan some wiring chases and some solid walls on the lower portion of the gazebo. You can use floor-mounted outlets, but they must be weatherproof outlets if they are used.

Many people like gazebos to have solid walls extended for a few feet from the floor. This eliminates the risk of furniture breaking out screen walls. The decision of what to do with walls will be up to your customers, but it is up to you to offer them options.

WALLS

When solid walls are used, you can finish them with any standard wall covering. However, if the upper walls are unprotected screens, you have to take moisture damage into consideration. If you use drywall under these conditions, make sure that it is moisture-resistant drywall. A better choice might be a wood interior finish that can be protected against weather degradation.

The upper portions of gazebo walls are likely to be made of screen. Some customers prefer windows that will open for good ventilation. Casement windows offer the most open area when windows are used. Installing windows in a gazebo is no different than installing them in a home.

Trade Tip: Ceiling fans are usually a big hit in gazebos. Many gazebos have vaulted ceilings and lend themselves well to fans. Even with a flat ceiling, you can use a close-couple fan to provide better circulation in the building.

Sliding glass doors are sometimes used to create the primary walls of a gazebo. There are drawbacks to this design. One of the major elements to consider is that the building will only have half of its wall space open. Glass will also make the gazebo hot during bright summer days.

The walls of most gazebos will be made of screen. When this is not the case, I favor a half wall and casement windows. You and your customers will ultimately determine what design to use.

ROOFING

The roofing for a gazebo is usually a small project. This doesn't mean it is simple. However, once you master the rafter cuts, the rest of the work is standard roofing. Seriously consider your rafter cuts and test a few of them before you pre-cut all of them.

SIDING

The siding used on a gazebo, when it is used, is the same type that would be used on any other type of residential structure. All of the options are open. There is no special requirement to worry about when siding a gazebo.

WINDOWS AND DOORS

Windows and doors for a gazebo can be nearly nonexistent. When a gazebo is screened, there is no need for windows. The door for this type of structure may be screened or solid. In any case, the door installation is standard procedure and requires no special consideration.

HEAT

Very few gazebos require heat. When heat is desired, there are two options. You will either connect to the existing heating system on the home where the gazebo will be built. Or, you will install an independent system. In most cases, an independent system will make more sense.

Electric heat can be used, but this gets expensive in extremely cold climates. A wall-mounted gas heater is a viable option. These units are available in models that don't have to be vented if they are not used in a bedroom. Natural gas is a suitable fuel, but if it is not available, propane gas will work with wall-mounted heaters. You have to know which type of gas you will be working with before you buy the heater.

Unvented gas heaters and electric heat are both inexpensive to install. Since gazebos are not used as full-time living space, the cost of operating heating units is of less concern than it would be for general living space.

PLUMBING

Plumbing is not usually installed in a gazebo. If it is wanted, you can usually bury piping in the ground to provide suitable plumbing. When the gazebo is to have plumbing and is located in a region where freezing temperatures are likely, the plumbing system must be protected from freezing.

If plumbing is installed in a gazebo, you must have some plan for concealing the piping. This can be something as simple as a corner chase to run the pipes in. Another option is separating sections of the walls with boxed columns that will conceal piping.

CEILINGS

The ceilings in a gazebo may be vaulted or flat. I prefer a vaulted ceiling. This type of ceiling offers a feeling of a larger

 Don't Do This! Don't attempt to use a gas appliance that is designed for use with natural gas when only propane gas is available. Likewise, don't use a propane heater when only natural gas is available.

space. Flat ceilings in small buildings make the usable space feel confined. This type of information should be given to your customers before they make their decision on which type of ceiling to use.

FLOORING

The flooring in many gazebos is simple decking. Indoor-outdoor carpeting is sometimes used. If carpet is to be used, be sure to install some sort of subflooring that will allow the tile to lay flat. I have seen carpet installed directly over decking, and the cracks between the decking can show through the carpeting. Tile can be installed as a floor covering, but if the tile become wet, it can be very slippery. An anti-slip tile is best when tile is used. Since gazebos are typically open to the weather at various times, choose a floor covering that will withstand wet conditions.

SCREENED PORCHES

Screened porches share many construction similarities with room additions and decks. There is not much that we have to talk about to cover screened porches that has not already been covered. The foundation work, floor framing, wall construction, and finish work is similar to the other projects that we have discussed.

One thing to keep in mind, just as you would with a room addition, is how the roof of the screened porch will attach to the home where it is being installed. Make sure that you have adequate height for the roof structure to connect to the home. If the porch will be a freestanding unit, this is not a consideration.

The advice I have given in this chapter for gazebos can be applied to screened porches. One thing to talk to your customers about is whether or not the porch will be screened completely. Some people with pets or children prefer the lower portion of their porches to be solid walls. Kids and pets have a way of damaging exposed screen. Aside from this consideration, your customers' desires will be the primary factor in what you build.

QUESTIONS AND ANSWERS

Q: When is a deck not worth building as a remodeler?

A: *I don't know that they are ever not worth building, but larger decks are more profitable than small decks. I tend to stick to decks that are at least 10 x 14 in size, but this is just my personal decision.*

Q: What do you think of building decks on patio blocks as a foundation?

A: *I don't like the idea. It is my opinion that decks should have a poured, pier foundation.*

Q: Can I let the base of deck steps land on raw ground if I use pressure-treated lumber?

A: *It is better to have a foundation under the terminal end of the steps and your local codes may require such support.*

Q: Is it really acceptable to build gazebos on skids?

A: *Yes. In fact, some customers prefer this construction design.*

Q: Why are gazebos so popular?

A: *They are attractive, affordable, and utilitarian.*

Q: Are screened porches better than decks?

A: *This is a personal opinion. Decks are great if you like to lay out in the sun. Screened porches are ideal if you want the feeling of being outdoors while being protected from insects. Both types of improvements have their place.*

Chapter 13

COSMETIC REMODELING

What is cosmetic remodeling? It is a term that is not well known among remodeling contractors. Think of cosmetic remodeling as small jobs that are affordable and that make a visual difference that is quite noticeable. Most remodeling has cosmetic elements involved, but cosmetic remodeling does not normally involve structural work. The purpose of cosmetic remodeling is to make buildings more attractive.

Homeowners often do their own cosmetic remodeling. But, there are many types of work that some homeowners are not capable of doing. And then there are those property owners who don't enjoy working with their hands and who hire contractors to perform simple tasks.

Most cosmetic remodeling involves a small amount of money when the jobs are compared to more traditional remodeling. For example, remodeling a kitchen can be very expensive, but refacing the cabinets in a kitchen is fairly affordable. Think in terms of little jobs that will make a big difference.

FOUNDATIONS

Foundations can be the subject of cosmetic remodeling. Homeowners are not always happy with the way the foundations of their homes look. Do you know how to spruce up a cinderblock foundation? What would you do to change the appearance of a pier foundation? Well, let's look at different types of foundations and see what you might want to do.

Cinderblock

Raw cinderblock foundations are abundant in many states. These are not the most attractive foundations in the world. Some people paint these foundations, but that doesn't add a lot of curb appeal to the foundation. There is, however, a way to transform a cinderblock foundation into a much more attractive base for a home.

The most effective method I have used to enhance cinderblock foundations is a simple one. Parge the foundation. Use patterns in the applications stage of the work. You can incorporate swirls into the parging process, or you can make the parged surface smooth. Once the mortar mix has been applied to the cinderblocks, you will have to wait for it to dry. Once it is dry, you can paint the foundation, usually with an earthtone color, to finish your transformation from a plain block foundation to an attractive base for a home. Throw in some foundation shrubbery and you have a winning foundation. This type of work doesn't cost a lot, but it makes a huge difference in the curb appeal of a home.

Pier Foundations

Pier foundations are not common in urban areas. This type of foundation is used more often in resorts and with vacation properties. Still, there are pier foundations found in many locations. What can be done to make this type of foundation more appealing?

Trade Tip: The foundations of buildings often settle in the first couple of years following the construction of the building. This can result in cracks. Even if the cracks don't pose a structural risk, they do attract attention and tend to give a bad impression of the quality of the building. If you have a brick foundation with cracks in the mortar joints, point them up and make the cracks go away. If you are dealing with a concrete foundation, seal the cracks with an epoxy-cement compound and once the repairs have dried, paint the foundation. Another option is to parge and paint the concrete foundation.

A fast, economical way to give a new look to a pier foundation is to simply install lattice panels between the piers. This doesn't take much time, and the lattice forms a barrier that conceals much of the underside of a home. You should use lattice that is rated for ground-contact use.

When a more expensive action is desired, you can close up the foundation with marine-grade plywood and install a brick façade on the plywood. This gets expensive and usually isn't worth the expense and trouble it takes to accomplish.

INTERIOR WALLS

Interior walls offer a plethora of opportunities for creative cosmetic remodeling. Many buildings have plain, painted walls that don't stand out. This is usually fine, but there are times when some accents on the walls would enhance a room. What can be done? You are limited only by your imagination when it comes to interior walls. Here are some ways to dress up interior walls:

- Stencil a design on the wall as a border where the wall meets a ceiling.
- Apply a wallpaper border where a wall meets a ceiling.
- Install crown molding where a wall meets a ceiling.
- Run wainscoting on the lower portion of a wall and finish it with a chair rail molding.

- Install wallpaper.
- Replace common electrical switch and outlet covers with designer covers.
- Texture a wall.
- Use ceramic tile to create accent areas on kitchen walls.
- Use weathered barn boards on the lower portion of a wall for a rustic look.
- Inlay a design into a wall with the use of natural wood boards.
- Install a mirrored panel on a wall.
- Build a shadowbox into a wall to hold collectibles.
- Recess a fish aquarium into a wall.

There is almost no end to what a creative mind can do to make interior walls interesting, functional, and attractive.

CEILINGS

Ceilings can give a creative remodeler some room to work. What can you think of doing to change the look of a ceiling without blowing a property owner's budget? Depending on how much money a person is willing to spend, there are many potential options. Here are some ceiling options:

- Texture a ceiling.
- Install false exposed beams for a rustic look.
- Consider installing a skylight or a light box with a skylight.
- Install a ceiling fan.
- Install recessed lighting.
- Hide cracked ceilings with ceiling tiles.
- Conceal ceiling defects with textured spackling and paint.
- Install a tongue-and-groove wood ceiling

FLOORING

It's not easy to do simple fixes on flooring. Carpeting offers no real options, short of deep cleaning. But, replacing the carpet in a small room isn't too expensive. You could replace sheet vinyl flooring with tile and make a substantial difference in the character of the room where the alteration is made. Generally speaking, there are no rabbits to pull out of the hat for flooring.

BATHROOMS

Bathrooms offer many possibilities for quick fixes that will not break the bank. Options can be as simple as replacing cheap, chrome towel bars with oak towel racks. Getting rid of an old soap dish and replacing it with a nice, brass or oak soap dish can make a difference. Lavatory faucets are not very expensive in the scheme of remodeling, and they are usually easy to replace. With a little thought, you can do a lot for a little in a bathroom.

- Replace a toilet.
- Replace a lavatory, a lavatory faucet, or a vanity top.
- Replace plain bathroom accessories with oak or brass accessories.
- Replace an old medicine cabinet with a new, fashionable cabinet.
- Replace a wall-hung lavatory with a vanity and molded lavatory top.
- Replace old flooring. Bathrooms tend to be small, so replacing a floor is not a major expense.
- Replace a shower curtain with a glass tub door.
- Wallpaper a bathroom.
- Stencil a boarder where the bathroom walls meet the ceiling.
- Refinish a stained bathtub.
- Regrout the tile in a bathroom.

KITCHENS

Kitchens, like bathrooms, offer a multitude of opportunities for cosmetic remodeling. From simple jobs, like replacing appliances, to more complicated projects, such as installing a garden window, kitchens provide plenty of opportunity for creative remodelers.

Cosmetic Remodeling Ideas for Kitchens

- ✔ Reface existing cabinets.
- ✔ Install a new countertop.
- ✔ Add track lighting in the kitchen.
- ✔ Install a garden window.
- ✔ Replace the flooring.
- ✔ Add a built-in dishwasher to the kitchen.
- ✔ Install an island cabinet and counter.
- ✔ Replace the old range hood.
- ✔ Install a replacement kitchen faucet.
- ✔ Replace the kitchen sink.
- ✔ Install replacement light fixtures.
- ✔ Install a tile backsplash along the countertop.
- ✔ Replace old appliances with new appliances.
- ✔ Install glass-front doors on cabinets.
- ✔ Install accent pieces to hold spices and other kitchen staples.
- ✔ Install under-cabinet lighting.

OTHER TYPES OF COSMETIC REMODELING

There are many other types of cosmetic remodeling. Very few remodelers think of landscaping as being part of their job, but it can be. An experienced remodeling contractor will look for all elements of what it takes to make a property stand out. This can include the addition of foundation shrubbery, a nice tree planted here and there, and so forth. You could extend this concept to include decorative flowerbeds, walkways, walkway lighting, accent lighting to showcase special portions of a building, stonework, and the list goes on.

Customers will sometimes know what they want when it comes to cosmetic remodeling, but it is often up to the contractor to educate property owners in the use of cosmetic remodeling. You can learn a lot about this concept by reading magazines. When you see an article that is interesting, save it and make notes. You will be able to use the information at some time in your remodeling work.

Trade Tip: Take photos of jobs that you do. Make a photo album to show perspective customers. Visual references go a long way in helping property owners get a grasp on your concepts.

IS IT WORTH IT?

Some contractors think that there is not enough money in cosmetic remodeling work. Is it worth it? Yes, it is absolutely worth your time and effort. There are two key ways to make money with cosmetic remodeling.

When you get a traditional remodeling job, look around and think about what you might offer the customer in the way of cosmetic remodeling. You are already on the job, so it's a great opportunity for you and your customer to reap the benefits of some small jobs. By selling the cosmetic services as an add-on to a larger job, you can increase your profit potential.

You can use cosmetic remodeling as a means of getting larger jobs. If you can get your foot in the door with something simple and inexpensive, you may be able to convince a property owner to expand the remodeling possibilities for their property. Go into a kitchen to replace the countertop, and leave with a full kitchen remodeling job. Or, go in for the counter replacement and leave with a contract to remodel a bathroom.

Cosmetic remodeling is not a phase of the business that you are likely to get rich with on its own, but it can add to your normal earnings and create new sales for larger jobs.

Trade Tip: What is the best way to get work? Word of mouth from satisfied customers is, by far, the best way to obtain new jobs. Use cosmetic remodeling to get into homes and to impress the homeowners with your professionalism. Even if the cosmetic customer isn't in the market for major remodeling, you may get a phone call from the customer's friend who wants a large project completed.

Cosmetic remodeling is a small niche in the industry, but it can result in many profitable jobs in years to come. Don't overlook this opportunity. Wise contractors expand their businesses in any viable way that they can.

QUESTIONS AND ANSWERS

Q: How much value does cosmetic remodeling have?

A: *Cosmetic remodeling is more about being happy than about making money. Most cosmetic remodeling is not the same type of solid investment that bathroom or kitchen remodeling is. On the other hand, cosmetic remodeling can sell a house faster than it would sell without the work. Ultimately, it is a personal preference on the part of the homeowner.*

Q: Will people really pay to have remodelers perform cosmetic remodeling?

A: *I think it would be difficult to make a specialty out of this type of work, but there are plenty of opportunities to capitalize on when you are already going to be working in a home.*

Q: Is it true that remodelers dealing in cosmetic remodeling need to be creative?

A: *It certainly helps.*

Q: I'm sorry, but I just think that most homeowners will do their own little household fix-ups. Is this not true?

A: *Clearly, many homeowners will do their personal touches around the house, but there are plenty of property owners who will pay a contractor to take care of the work.*

Chapter 14

DUST AND DEBRIS CONTROL

Controlling dust and debris on a remodeling job is an important element of the job. It seems simple enough, but it can be a real problem. Sanding drywall mud, as an example, is a very dusty job. Homeowners don't take kindly to having their entire home covered in the dust created by the installation of new drywall. Sawdust is another airborne invader that finds its way into living space. If you don't want to deal with irate customers, you have to implement satisfactory procedures for the control of dust and debris.

You don't have to be a rocket scientist to come up with viable ways to limit the invasion of dust and debris into the living space of homes and buildings where your crews are working. However, if you don't take an active role on this issue, you are surely going to receive numerous complaints from your customers.

Trade Tip: Plastic sheeting and duct tape can go a long way in preventing the migration of dust from a work area into a living area.

MESSY JOB SITES

Messy job sites usually are not appreciated by customers, contractors, or crews. You should have a company policy that demands a clean and organized job site. Not only are messy job sites unpleasant, they are not safe. How many times have you seen a block of wood with a nail driven through it laying on a floor? I don't know about you, but I have seen this type of situation far too many times. Even with good boot on, a person's feet are not protected from a nail that is pointed upward and secured in a block of wood. This should concern you for your own safety. If this is not enough, think of the safety of your crews and subcontractors. Take it to another level and image a customer walking into the work area and stepping on the nail. If you really want to think hard, imagine a child stepping on the nail. It's an ugly scene. Can you afford the lost time from work that an infected foot will give you? Are you willing to face the lawsuit that might well be filed against you when someone else steps on the nail? Why take this risk? Insist that your workers pick up such hazards and neutralize them.

Pieces of wood with nails sticking out of them are not the only risks to be aware of. If you are working on a job and leave electrical tools plugged in and unsupervised, you could find yourself in a very bad situation if a child picks up a circular saw and pulls the trigger out of curiosity. Don't leave ladders set up when they are not in use. Cordon off work areas to keep people out of the site who don't have a legitimate reason for being there. What does this have to do with dust and debris control? The wood block with the nail would be considered debris. Think of the rest of this advice as words from the wise to keep you out of trouble and out of court.

What are some other types of job-site debris that pose a safety hazard? Scrap wood can create a tripping hazard. Used razor blades left on an countertop can result in serious cuts. Broken glass is an obvious hazard. The list could go on and on. Make sure that your crews don't leave you exposed to lawsuits.

Require your crews to clean up hazardous situations when they are created. We all know that remodeling results in a lot of debris that can be dangerous. It's not uncommon for crews to work around the hazards and either leave them or clean them up at the end of the day. This isn't good enough. Let your crews know that if you, or your field supervisor find hazards during unexpected visits to the job site that there will be serious consequences for the person in charge of the job site. Take me seriously on this issue.

DEBRIS REMOVAL

What is your plan for debris removal? Will debris be removed by truck on a daily basis, or will you have a portable trash receptor put on the job site? The decision of how to handle debris removal will generally depend on the nature of the job. If you are doing a small job, hauling the debris away each day is probably the correct answer. A large job, especially one that requires extensive rip-out work, is worth putting a portable trash container on the site. You should build the cost of this service into your bid.

Some property owners will complain about having a trash receptacle placed in their driveway or on their lawn. Cover this issue up front, before you take the job. If you have to truck large quantities of debris from a job, the expense will be much higher than having an on-site trash receptacle.

Avoid creating trash piles that will remain on the job site at the end of the day. This was once a common practice, but piling trash, lumber, and other debris on a job site creates several concerns. One concern is the potential fire risk of having such a pile close to a building. Another risk is that children will play on the pile and become injured. Good advice is to avoid trash piles.

 Don't Do This! Don't ignore safety hazards on your jobs. You should be insured, but insurance is not an excuse for making stupid mistakes and leaving traps for innocent people to fall prey to. If your actions are negligent, insurance may not cover your claim.

EXISTING FLOORS

One responsibility when remodeling is the protection of existing floors. This includes flooring that must be crossed to get to the work area and, on occasion, the flooring in the work area. Floors are usually protected with heavy paper that comes on a roll, tarps, or dropcloths. Any of these types of protection can work, but some work better than others for specific purposes.

Rolled paper does a very good job of protecting floors. When installed tightly, the paper is not slippery and has a long life during the course of a job. Tarps tend to be a tripping hazard and they don't always contain liquids well. Canvas dropcloths work very well in protecting floors of work areas. Cardboard is also used to protect floors. Any floor protection used should be secured firmly to avoid tripping people.

Characteristics of Materials Used for Floor Protection

- ✔ Rolled paper works very well for creating walkways to work areas and for covering the floors of work areas.
- ✔ Tarps are waterproof, but can present a tripping hazard if not stretched tight and secured.
- ✔ Canvas dropcloths are great for floor protection when painting.
- ✔ Plastic dropcloths do not perform well as floor protection.
- ✔ Foam padding, like that used as a carpet pad, is a terrific floor protector.
- ✔ Cardboard, when secured firmly, provides good floor protection.

DUST

Dust is a major problem during remodeling. It's nearly impossible to control all of the dust created, but there are steps that you can take to limit the amount of dust leaving the work area. Where extreme dust is present, workers should wear adequate protection equipment, such as a dust mask, to protect them from the potential effects of breathing in the dust.

There are three major categories of dust for remodelers to deal with. The first type is general dust that is created during a workday. Dust created when sanding drywall mud is plentiful, white, and nasty if it gets tracked through a home or building. Fortunately, the dust from joint compound is fairly heavy and tends to fall to the floor near where the sanding is taking place. This makes it easier to contain the dust. Sawdust is the third type of dust that is found in large quantities on many types of remodeling jobs. Once you contain these three types of dust, you have most of the battle won. Here are tips for controlling general dust:

- Install suitable floor protection that will allow easy vacuuming.
- Keep windows closed to avoid circulating dust to areas outside of the work space
- Choose one door that will be the primary door used during the job.
- Close all doors that will not be needed and seal them with plastic and duct tape. To do this, cut sheets of plastic larger than the doorframe and tape the plastic to either the doorframe or wall. You can substitute masking tape for duct tape if you are concerned that the adhesive on the duct tape will create problems.
- The door that will be used regularly cannot be sealed completely. Use plastic and tape to secure the upper portion of the door and one side of the plastic. Wrap the bottom of the plastic about a length of wood and staple the plastic to the board. By doing this, the weight of the wood will hold the plastic down and still allow ingress and egress to and from the work space.
- Keep a vacuum on the site and use it often.

To control dust created from joint compound, vacuum frequently. Use all of the methods recommended for general dust, and don't allow workers to track drywall dust into other areas. This dust sticks to skin, clothes, and boots. Avoid tracking it by staying out of other areas. If you must pass through other space, use a hand vacuum to get the bulk of the dust off of clothes and footwear before leaving the work area.

Sawdust is always present on most remodeling jobs. One of the best ways to keep sawdust out of living space is to set up a cut area. Designate an area outside of a building, in a garage, or some similar place to contain sawdust. If you can keep the saws out of the building, you will eliminate most of the sawdust-related problems.

Trade Tip: A battery-powered, handheld vacuum is a good tool to keep on a job. Workers who won't go to the trouble of using a large vacuum for a small pile of dust will grab a ready-to-go handheld vacuum. Any dust that you can vacuum in the work area is dust that you won't have to worry about in other parts of the building.

CLEAN JOB SITES

Clean job sites are more productive. Property owners appreciate a contractor's effort to keep a job as clean as possible. Controlling dust and debris will not cost a lot of money or take a lot of time, but it will make your jobs safer, cleaner, and probably more successful.

QUESTIONS AND ANSWERS

Q: It seems to me like the homeowner should deal with dust from a job. Why is this wrong?

A: *The contractor is being paid to do a job. Unless there is some special consideration in the contract between a contractor and a customer, the contractor is charged with the responsibility of all aspects of the job. This includes any clean-up work.*

Q: How often should trash receptacles be dumped?

A: *I think it depends on the situation. It is common to wait until the containers are nearly full to order a pick-up. However, if there is reason to believe that children may get into the trash containers during a weekend, I think that the containers should be dumped at the end of each work week.*

Q: Do you prefer contract debris removal, or do you prefer to handle it yourself?

A: *I prefer contracted services for debris removal.*

Q: How do you feel about burning debris piles?

A: *This practice is usually a violation of law. Even if it is not, I would not approve of the procedure.*

Chapter 15

RIP-OUTS OF EXISTING CONDITIONS

Rip-outs of existing conditions can be the best part of a remodeling job. If you are harboring frustration, ripping out walls and busting up bathtubs can free you from the weight that is on your shoulders. Seriously, it can feel good to tear things up from time to time. But in reality, most rip-outs are done in a controlled, orderly fashion.

Bathroom and kitchen remodeling always involves some degree of rip-out. Other types of remodeling can also involve removing existing materials. Even if all you are doing is replacing windows in a home, you have to take out the old windows before you can install the new windows.

How hard can it be to tear things out of a building? Well, it can get tricky, and there are some ways that are smarter and more productive than others. Experience plays a big factor in making a rip-out profitable. Rookie remodelers can waste a lot of time and bust their backs on jobs where seasoned remodelers would use different methods and be done faster and without the backaches.

When you are doing an aggressive rip-out, you have to think about more than just the workspace. Consider a job where you are on the second floor of a home and are trying to remove a tile wall and a cast-iron bathtub. If you take a sledgehammer to the tub, you may crack a ceiling in a room on the first floor. When this happens, you have to fix the ceiling, and there goes a chunk of your anticipated profit. There are certainly tricks of the trade when it comes to rip-outs.

Rip-outs vary from job to job. It's a lot easier to rip out a living room than it is to rip out a kitchen. In a living room there is not much to have to tear out. With this concept in mind, let's do something of a room-by-room look at rip-out work.

GENERAL LIVING SPACE

General living space does not present many rip-out problems. What is general living space? For our purposes here, let's consider it to be living rooms, family rooms, dining rooms, bedrooms, and similar spaces. Basically, the topic here is a room that doesn't have plumbing, cabinets, or other substantial elements to be removed.

If you are going to give a room a full facelift, you may have to remove the flooring. This is not difficult if the flooring is carpet. Rooms with wood floors can either have the floors refinished or covered with carpet. Removing carpeting is simple, but heavy work. You must be prepared to remove the old carpet, and possibly the pad, from the job site.

Light fixtures will probably be removed and replaced with new lights. Again, this is a simple procedure. It's possible that you will replace windows or install new windows or skylights, but this is not a challenging rip-out. All in all, general living space doesn't require a lot of effort or thought to prepare for remodeling.

 Trade Tip: Make eye protection available to all members of your rip-out crew and insist that it be worn when doing demolition work.

Trade Tip: If you will be removing electrical fixtures, make sure the wires that are left for later reconnection are safe and secure. Install wire nuts on the wires, tape the nuts to the wires, and seal the opening of the electrical box to keep people from coming into contact with the wiring.

BASEMENTS

Basements are often finished into living space. After awhile, they are sometimes remodeled. Generally, the biggest problem with remodeling a basement is ingress and egress with supplies and materials. Rip-outs in basements are not much different than ripping out general living space.

ATTICS

Attics are sometimes converted to living space. This can, and often does, involve the installation of dormers. What is there to rip out of an attic? There can be plenty. For example, the ceiling joists for the room below the attic may not be large enough to support living space in the attic. Collar ties might have to be moved around. There are even times when the entire roof structure is removed and replaced with a new structure that will provide more living space. This is a major rip-out.

When beefed up joists are needed to support new living space, existing joists are usually left in place and reinforced with the installation of new joists. Sheets of plywood or wafer board that has been installed as walkways in an attic are generally removed.

Don't Do This! It can be tempting to cut a hole in a roof or gable wall to throw trash and debris out of during an attic conversion. Don't do it. Someone may get hurt on the ground. Either put someone on the ground to keep people out of the danger area, or used a chute to direct debris into a portable trash receptacle.

Due to limited access to attics, some provisions are needed to allow sheets of subflooring and drywall to get into the attic. This is sometimes done by cutting a hole in the gable end of the attic for access that will later be filled with a window. Other than protecting people on the ground while this type of hole is being cut, the job is not difficult. The same concept applies to making roof cuts for a dormer. You have to secure the area below the work area. This can be done with *caution tape*. However, it is best to have a crewmember on the ground, in a safe location, to make sure that no one comes under the work area at the wrong time.

BATHROOMS

Ripping out bathrooms is a common occurrence for remodelers. Bathroom and kitchen remodeling accounts for a high percentage of all remodeling work done. Unlike general living space, bathrooms have plumbing fixtures to remove and often require the removal of tub surrounds. It's not uncommon to have the finished flooring in a bathroom removed. Experienced remodelers can make fairly short work of a bathroom rip-out, but the work does require some extra consideration. Here are suggestion for bathroom rip-outs:

- Don't assume that plumbing valves will be in working order. Check the valves to see that they do, in fact, cut the water off completely.
- If there is only one bathroom, plan provisions for sanitary services for the period of time that the existing bathroom is out of commission. This can require the rental of a portable toilet compartment or the removal and reinstallation of the existing toilet until such time as the new toilet can be installed.
- Be aware that the water supply to the bathing unit is not required to have individual cut-off valves, so the entire water supply will have to be turned off to allow for the removal of a tub or shower faucet.
- Wear eye protection when removing ceramic tile from floors and walls.
- Caution is needed during heavy demolition to prevent cracks in nearby walls and ceilings outside of the work area.

- Check flooring for potential water rot before recovering the floor surface.
- A plumbing permit should not be required if you are replacing existing fixtures in their exact, previous location.

Bathroom Floors

Bathroom floors are often replaced. Sometimes tile floors are replaced with vinyl flooring. At other times, vinyl flooring is replaced with tile. When a bathroom floor is stripped down to the subflooring, the floor structure should be inspected for potential rotting spots. If you are charged with removing a tile floor, you must be careful not to damage a ceiling that is below the bathroom, if there is a ceiling below the bathroom. Hammers and chisels are common tools for tile removal. Banging on the floor can cause globes from light fixtures to fall. Remove the globes prior to the demolition work. Wear eye protection to avoid chips of tile from creating eye injuries.

Tile Walls

Tile walls are often removed during a bathroom makeover. There are a lot of old bathrooms that have tile installed about halfway up the wall. Many homeowners detest this look. It is often possible to strip the tile from the drywall, without damaging the wall structure. But, what's left of the drywall is rough and often unusable. I favor removing the drywall with the tile still attached to it. This is faster and doesn't create as much mess. By installing new, water-resistant drywall, you are giving the customer a better wall surface for painting, wallpapering, or tiling.

Tile Tub Surrounds

Tile tub surrounds were very popular for many years. Today, many homeowners prefer fiberglass tub surrounds. When

Trade Tip: If you are going to install wallpaper in a bathroom, make sure that the type of wallpaper to be installed is suitable for high-moisture applications.

remodeling a bathroom, it is likely you will be asked to remove a tile tub surround. If you are going to replace the tile with a sectional, fiberglass surround, you can use a chisel to remove the tile. This should leave the drywall in good enough shape for you to install a fiberglass surround on it. If the tile will not strip off of the drywall, cut it out and replace the drywall.

Bathtub Removal

Bathtub removal during a rip-out can be quite a job, especially if the tub is made of cast-iron. Many older bathtubs are made with cast-iron. These puppies are heavy, pushing over 400 pounds at times. A tub like this can be removed in one piece, but the effort required is substantial. Unless the tub has value, such as in the case of a claw-foot tub, I generally prefer to break the tub into small pieces for removal. This is done with a sledgehammer while wearing eye protection. Banging on a cast-iron tub with a sledgehammer is going to rock a home. Make sure that pictures are removed from walls, globes on light fixtures are removed, and that you don't crack walls or ceilings out of the work space during the demolition.

Shower Bases

Shower bases are easier to remove than bathtubs are. Once the wall material coming down over the flange of the shower base is cut out, disconnecting the base from its drain is all that is left to do.

Vanities and Lavatories

Vanities and lavatories are normally small enough that removal is not a problem. Make sure that all water supplies are turned off

Trade Tip: Wear heavy, protective gloves when handling pieces of cast-iron fixtures. Shards of the enamel will cut unprotected skin quickly and can become embedded in the skin. It is also wise to wear a long-sleeved shirt when breaking up a cast-iron fixture to avoid flying shards from finding exposed skin.

Did You Know: that toilets have integral traps that retain water even after a toilet is removed from a closet flange? If you attempt to carry a toilet bowl out of a building, you can expect to have a significant water spill somewhere along the way. Here's how you avoid that. Pull the toilet before you demo the bathing unit. Place the toilet in the bathing unit and move it into various positions until all of the water has drained from it. If a spill is still concerning you, stuff the drain hole of the toilet with an old towel or some other item that will retain water and that can be disposed of along with the used toilet bowl.

and be careful not to damage plumbing pipes when removing these fixtures. If the fixture will not be replaced during the same workday, cap the drainpipe to prevent sewer gas from entering living space. A neoprene cap with a stainless-steel clamp is easy to install and extremely effective in blocking the escape of sewer gas.

Toilets

When removing toilets, make sure that the water supply is turned off. Wear eye protection. Most toilets are made of china and they do explode if they are put under stress. If this happens, sharp pieces of china go flying about the work area. You may have to cut the bolts that secure the toilet bowl to the closet flange.

Bidets

You can treat the removal of bidets with the same cautions that apply to the removal of toilets. Bidets and toilets are very similar in nature and can be dealt with using the same basic procedures.

Light Fixtures

Light fixtures, medicine cabinets, towel racks, and other items that must be removed for a full remodeling job on a bathroom are easy to eliminate and don't require any special rip-out advice.

KITCHEN REMODELING

Kitchen remodeling is more complicated than general remodeling. There is a lot of material to be removed from a kitchen before a new one can be created. I've always thought that kitchen rip-outs are easier than bathrooms in some ways. There is more room to work in a kitchen than there is in a bathroom. You don't have to monkey around with 400-pound bathtubs in a kitchen. Most kitchens are on ground level, where many bathrooms are on upper floors. But, there is more to come out of a kitchen.

Flooring, Walls, and Ceilings

The rip-out of flooring, walls, and ceilings for kitchen remodeling is comparable to the working conditions in a bathroom. There are no special considerations for these areas that we have not already covered.

Cabinets and Counters

Cabinets are a large part of a kitchen. This includes base cabinets and wall cabinets. It is not uncommon for existing cabinets to be removed for replacement with new cabinets. Sometimes existing cabinets are refaced to save money. There are two sets of conditions under which cabinets are removed. One is when they will be saved for reuse and the other is when they will be scrapped. If they are going to the dump, how you take them out doesn't matter much. But, if you are going to reinstall the cabinets, you must remove and store them with care.

Counters are frequently replaced when remodeling a kitchen. Sometimes the counters have to be cut for removal. There's nothing really tricky about getting rid of an existing counter. If

Did You Know: that kitchen and bathroom remodeling provides the highest rate of return on investment for homeowners? Well, most any real estate appraiser you talk to will attest to the value of remodeling kitchens and bathrooms. This is a good selling point for contractors when talking with potential customers.

there is a sink in the counter, you will have to disconnect the plumbing prior to removal. You can use the same basic rules for lavatory removal to rip-out a kitchen sink. If a garbage disposer is connected to the sink, you will have to disconnect the electrical wiring from it and protect the bare wires from coming into contact with people.

Appliances

Appliances in kitchens include refrigerators and ranges. Dishwashers are common kitchen appliances. Trash compactors and other appliances may come into play. When you remove a refrigerator, be careful not to break or crimp the water line that delivers a water supply to an icemaker. You can look in the freezer of the refrigerator to see if an icemaker is in use.

There will be both a water supply and a drain to disconnect when a dishwasher is going to be removed. The drain hose can be cut and plugged. As long as the valve on the water supply is functioning properly, disconnecting the water supply is no big deal. You will also have to disconnect the electrical wiring. Make sure to install wire nuts on exposed wiring, tape the nuts in place, and secure the wiring in a safe place.

Some contractors keep kitchen appliances in place for as long as possible to avoid disruption for homeowners. This complicates the remodeling procedure, but it is sometimes necessary. When you rip-out appliances, you must make sure that they are safe for storage until they are disposed of. In the case of a refrigerator, this means that you should remove the doors on the appliance to prevent children from becoming trapped.

Doing rip-out work is not for everyone, but it can actually be fun. Demolition work is a given part of many remodeling jobs. Remember that the demo work is done before the finish work, so keep your customers happy during the destruction of their property. It will make the end result easier to reach.

QUESTIONS AND ANSWERS

Q: Should I send apprentices in to do a rip-out?

A: *Not alone. Rip-out work requires a combination of skill and strength. Don't trust green rookies to handle a rip-out on their own.*

Q: When should I begin a rip-out?

A: *Don't tear out needed amenities, such as a kitchen sink or a toilet, until you are ready to replace them quickly.*

Q: How do I minimize the effects of a rip-out?

A: *Establish a containment procedure to control dust, noise and debris. Try to work when your customers are not on the site. This is about the best that you can do.*

Q: How long does an average bathroom rip-out take?

A: *An average crew of two can tear a bathroom apart in a day.*

Q: What amount of time is needed to gut a kitchen?

A: *An experienced crew of two or three workers can take out an average kitchen in a day.*

Q: Can the carpet in an average home be removed in a single day?

A: *Yes.*

Q: Is it wise to let homeowners do their own rip-out work?

A: *Usually it is not wise to allow homeowners to do their own rip-outs. Of course, if the homeowner insists, there is not a lot that you can do about it. However, if you agree to allow this type of situation, reserve the right to change your bid price if the rip-out is not done properly.*

Chapter 16

SAFETY ISSUES

Safety should be a primary concern on the job. Anyone who has worked large jobs knows that OSHA regulations are numerous and that fines for violating the rules can be substantial. Some people don't like the rules, but they are designed to protect workers. Yet, you will find few such regulations on small, residential jobs. Does this mean that you are less likely to be injured on a residential job? I suppose you could say it is a numbers game, but believe me, you can get hurt building a house as quickly as you can building a skyscraper (Figure 16.1).

Think about the jobs that you have worked on recently. How many single-family home job sites even have a portable toilet on the site? Do you remember seeing hard hats being worn on the job? I doubt it. How long has it been since you have seen a long electrical extension cord wrapped with duct tape as a substitution for cut insulation on the cord? I would guess that this image is fresh in your mind. Let's face it, small residential jobs are not seriously regulated for safety procedures. Will a roll of 15 pound roofing felt coming off the roof of a two-story house and hitting your bare head hurt you any less than it would coming off a two-story commercial building? I don't think you would notice the difference. Yet, the regulations are much more lax on residential jobs.

I am not suggesting that the trades need more regulation. It should be the goal of every contractor and worker to practice safe working procedures without being told to. If that roll of roofing felt hits you on the head, are you going to feel any better that you are fined for not wearing a hard hat? Do you think that your head will appreciate the fact that the person responsible for the felt hitting you was not in compliance with required safety procedures. If you are into lawsuits, you might like it, but if you value your head and your safety, the best answer is to avoid getting hurt in the first place.

Have you heard how you are a wimp for using safety equipment? Join the crowd. I don't know of anyone who has been in the trades for a significant period of time who has not been ribbed about using safety equipment. Personally, I have been teased about wearing safety glasses when breaking concrete. People have made fun of me for wearing ear protection when running a jackhammer. Let them laugh. I can still see and hear. Admittedly, I was influenced, to some extent, by the cajoling when I was a rookie. I regret that now that I am older and suffer

---

General Safe Working Habits

1. Wear safety equipment.
2. Observe all safety rules at the particular location.
3. Be aware of any potential dangers in the specific situation.
4. Keep tools in good condition.

FIGURE 16.1 General safe working habits.

Trade Tip: Keeping your tools in good condition is a wonderful way to make your work more productive, more profitable, and safer. Keep drill bits and saw blades sharp. Check electrical cords every day to see that they are safe. Inspect your equipment and use it for the purpose it is intended to be used for.

some signs of what might have been prevented if I had not succumbed to peer pressure. As a seasoned professional, I can tell you to use your safety gear from day one. You are not a wimp. Being responsible about your safety and the safety of others makes you intelligent. The people who rag on you are the idiots.

CLOTHING AND JEWELRY

Clothing and jewelry are common human items. Clothing is mandatory in most cases. But, what you wear on the job site can get you hurt or keep you safe. Knowing what to wear and how to wear it is an important step in job safety (Figure 16.2).

Jewelry has a very limited purpose on a job site. Lots of people wear rings, chains, and other personal items. I suggest you leave the jewelry at home. The potential injuries associate with jewelry can be serious.

I used to wear a wedding ring to work. On one occasion, I was driving a steel post into the ground. With one swing of the sledgehammer, I cut my ring finger to the bone. A rib on the post hooked into the ring of my left hand that was holding the post and created a nasty cut. This would not have happened if I had not been wearing the ring. From that day forward, I have not worn a ring while working on a job. Anything worn around your neck could result in a bad injury. I could go on with war stories, but I assume you get the idea.

Now let's talk about clothing. Loose clothing and shirttails that are not tucked in can be a big problem. Here's another war story for you. I was on one knee drilling a large hole in the floor of a bathroom one day. The drill was a right-angle drill with a mean bit in it. The bit hit a knot or a nail and kicked up. To my surprise, the drill was locked in the on position. I had not meant

---

Safe Dressing Habits

1. Do not wear clothing that can be ignited easily.
2. Do not wear loose clothing, wide sleeves, ties or jewelry (bracelets, necklaces) that can become caught in a tool or otherwise interfere with work. This caution is especially important when working with electrical machinery.
3. Wear gloves to handle hot or cold pipes and fittings.
4. Wear heavy-duty boots. Avoid wearing sneakers on the job. Nails can easily penetrate sneakers and cause a serious injury (especially if the nail is rusty).
5. Always tighten shoelaces. Loose shoelaces can easily cause you to fall, possibly leading to injury to yourself or other workers.
6. Wear a hard hat on major construction sites to protect the head from falling objects.

FIGURE 16.2 Safe dressing habits.

to lock it on, but I had. The bit jumped from the floor to my exposed shirttail. In short order, the bit crawled quickly up my shit, ripping off buttons as it went. By the time I could stop it, the worm of the bit was in my right arm and it was all I could do to keep the sharp teeth from tearing into my arm and armpit. A helper unplugged the drill, but not until I had suffered a painful and potentially serious wound. A trip to the hospital showed that if the bit had been about an inch farther in one direction I would have lost the use of my right arm. This is not a good thing, and it was all due to having my shirttail out.

TOOLS

Tools account for many accidents. Saws are particularly nasty when they get out of control. Large drills can do a lot of damage. Even a simple hammer can crush bones quickly. The improper use of ladders and scaffolding often results in serious injuries. Some tools are given more respect that others. A simple screwdriver can put an ugly hole in your hand if it slips during use. My point is this: any tool can take a toll on your safety (Figure 16.3).

There are countless stories of injuries related to the misuse of tools. These types of injuries range from electrocution to broken bones, and worse. Most accidents are avoidable. Many accidents occur when workers are trying to save time, take shortcuts, or not thinking about what they are doing. Leaving a hammer on a high ladder and climbing down to have the hammer land on your head is one example, but it is only one example of some much more serous injuries that could occur.

There are all types of tools. Since you are a professional, I will not bore you with a full listing of them. It is often the improper use of hand tools that results in injuries on the job site. Tools as simple as a cold chisel and hammer can blind you if you are not wearing eye protection when using them. Don't be stupid. Use your safety gear and common sense when using hand tools (Figure 16.4).

Don't Do This! Don't wear sneakers on a job site. Boots with heavy soles provide a lot more protection from nails and other sharp items that can injure your feet. I see people with sneakers constantly. They may be fashionable, but they are not practical for most job-site applications.

Safe Operation of Grinders

1. Read the operating instructions before starting to use the grinder.
2. Do not wear any loose clothing or jewelry.
3. Wear safety glasses or goggles.
4. Do not wear gloves while using the machine.
5. Shut the machine off promptly when you are finished using it.

FIGURE 16.3 Safe operation of grinders.

> **Safe Use of Hand Tools**
>
> 1. Use the right tool for the job.
> 2. Read any instructions that come with the tool unless you are thoroughly familiar with its use.
> 3. Wipe and clean all tools after each use. If any other cleaning is necessary, do it periodically.
> 4. Keep tools in good condition. Chisels should be kept sharp and any mushroomed heads kept ground smooth; saw blades should be kept sharp; pipe wrenches should be kept free of debris and the teeth kept clean; etc.
> 5. Do not carry small tools in your pocket, especially when working on a ladder or scaffolding. If you should fall, the tools might penetrate your body and cause serious injury.

FIGURE 16.4 Safe use of hand tools.

Electrical tools are often very powerful and potentially dangerous (Figure 16.5). Even so, many workers take them for granted. From my experience, most injuries with these tools occur with two groups of workers. Rookies are the most likely workers to fall prey to electrical tools. Old pros who have become complacent with their tools are the second group. People in the middle of these two extremes don't seem to get hurt as often. I don't have statistics to back this up. My representation is based on my personal observation on jobs over my 25 year career in the trades. It seems to me that young workers don't know what the tools can do to them. Older workers feel like they have mastered the tools and won't get hurt. Anyone can be hurt at any time. Be alert!

Tools are not the only source of injuries on jobs. Whether you are working high or low, there are risks involved. I have had ladders fall with me on them and I've been partially buried in trenches. In both cases, the fault didn't seem to be mine, but there were probably steps I could have taken that would have reduced the risks. Realistically, I would have been covered in dirt either way. It's easy to say do this and do that, but when

Safe Use of Electric Tools

1. Always use a three-prong plug with an electric tool.
2. Read all instructions concerning the use of the tool (unless you are thoroughly familiar with its use).
3. Make sure that all electrical equipment is properly grounded. Ground fault circuit interrupters (GFCI) are required by OSHA regulations in many situations.
4. Use proper-sized extension cords. (Undersized wires can burn out a motor, cause damage to the equipment, and present a hazardous situation.
5. Never run an extension cord through water or through any area where it can be cut, kinked, or run over by machinery.
6. Always hook up an extension cord to the equipment and then plug it into the main electrical outlet—not vice versa.
7. Coil up and store extension cords in a dry area.

FIGURE 16.5 Safe use of electric tools.

Trade Tip: Don't leave electrical tools connected to a power source when you are not using them. Imagine a scene where a homeowner comes to a job to inspect the progress of the job. Assume that this homeowner has a young child along for the visit. If the child picks up a circular saw that is connected to electrical power and pulls the trigger, the results may not be something that you would want to live with for the rest of your life. Safety on the job is not only about you.

you are working in the real world, a safety lecture doesn't always play out.

Let's start from the bottom and work our way up. People who work in trenches and ditches are at risk from a number of factors (Figure 16.6).

Don't Do This! Don't use tools for anything that they are not intended to be used for. As an example, don't use a screwdriver as a chisel or as a pry tool. Many injuries occur when tools are used for purposes other than for what they were designed.

DITCHES AND TRENCHES

Ditches and trenches claim lives every year. It's common for special safety equipment to be used to protect workers in many types of trenches. But, a lot of people work in ditches and trenches where shoring equipment is not used. There are multiple risks involved with working below grade level. Even a relatively shallow ditch can be extremely dangerous.

I was working in a shallow ditch one day many years ago. It was a sewer ditch that was only about five feet deep. There was a large mound of dirt piled up on the edge of the ditch. While I was inspecting the connection between a building sewer and the sewer, I heard equipment in the area. It was a backhoe. There was no reason for me to feel uncomfortable as I leaned over the connection. The backhoe had been passing the area many times without incident. Little did I know that this time would be different.

For some reason, the backhoe operator decided to spin his piece of equipment around. The back bucket and arm on the backhoe hit the pile of dirt next to the ditch. By the time I realized what was happening and tried to move, I was buried from my waist to my feet with a heavy layer of dirt. While I was not injured, I could not just stand up. Fortunately, I had a shovel in the ditch to dig myself out with and there were no additional loads of dirt coming into the ditch. The backhoe operator didn't know I was in the ditch or that I had been trapped. I dug myself out. Aside from being a bit frightened and very angry, I was not hurt. However, this accident could have proved fatal under other circumstances. It was an accident.

What could I have done? Should I have placed a person on ground level to protect me? That would have been a good idea, but who has the budget for that type of security? Should the backhoe operator stopped to see what had happened? Maybe,

---

Rules for Working Safely in Ditches or Trenches

1. Be careful of underground utilities when digging.
2. Do not allow people to stand on the top edge of a ditch while workers are in the ditch.
3. Shore all trenches deeper than 4 feet.
4. When digging a trench, be sure to throw the dirt away from the ditch walls (2 feet or more).
5. Be careful to see that no water gets into the trench. Be especially careful in areas with a high water table. Water in a trench can easily undermine the trench walls and lead to a cave-in.
6. Never work in a trench alone.
7. Always have someone nearby—someone who can help you and locate additional help.
8. Always keep a ladder nearby so you can exit the trench quickly if need be.
9. Be watchful at all times. Be aware of any potentially dangerous situations. Remember, even heavy truck traffic nearby can cause a cave-in.

FIGURE 16.6 Rules for working safely in ditches or trenches.

but we all know how likely that would have been. The simple inspection of a plumbing connection could have resulted in my death. Don't take ditches and trenches lightly.

LADDERS AND SCAFFOLDS

Ladders and scaffolds are responsible for many injuries. Some of the injuries are what you would expect. Others are not. I once tore my stomach muscles while drilling floor joists on a ladder because I reached too far to avoid having to move the ladder. On

Did You Know: that ditches that are deeper than 4 feet should be shored up? How often does this happen on small jobs? Not often enough. Be careful when you are below grade.

another occasion, I rode an extension ladder across the side of a commercial building on a damp morning when a gust of wind hit me. There was also a time when a piece of motorized equipment ran into a rolling scaffold I was on and I was nearly knocked to a concrete floor some 20 feet below. I know a painter who broke his legs in seven places when his extension ladder collapsed. There are a host of injuries possible when working from elevated positions (Figure 16.7).

There are some common sense elements to using ladders. Don't use an aluminum ladder when working with electricity. Wood and fiberglass ladders offer a greater degree of safety when working with electricity. Never climb to the top of a stepladder. Those warning labels that you will find on the top rung of a stepladder are there for a good reason. Make sure the base of any ladder is on a stable, level platform. Check the weight rating of any ladder you are about to use. Many ladders are rated for individuals of weights that are less than some potential users. Workers must match their body weight with suitable ladders. When setting up an extension ladder, pay attention to the angle of use that the ladder is rated for. Like any tool, use ladders for their intended purposes. How many times have you seen people lay boards between two ladders for use as a work platform? I would guess you have seen it too many times. Don't do this. There is scaffold equipment available for walk boards. You don't have to use ladders for this purpose.

Don't Do This! Don't leave tools perched on the top of tall ladders. Even an 8 foot stepladder is tall enough for a falling tool to do some serious damage to a human body. It's far too easy for a ladder to be bumped into so that a tool left on it will come tumbling down. I know it is tempting to leave the tool there for just a moment, but don't do it.

Safety Issues **347**

> **Working Safely on a Ladder**
>
> 1. Use a solid and level footing to set up the ladder.
> 2. Use a ladder in good condition; do not use one that needs repair.
> 3. Be sure step ladders are opened fully and locked.
> 4. When using an extension ladder, place it at least ¼ of its length away from the base of the building.
> 5. Tie an extension ladder to the building or other support to prevent it from falling or blowing down in high winds.
> 6. Extend a ladder at least 3 feet over the roof line.
> 7. Keep both hands free when climbing a ladder.
> 8. Do not carry tools in your pocket when climbing a ladder. (If you fall, the tools could cut into you and cause serious injury.)
> 9. Use the ladder the way it should be used. For example, do not allow two people on a ladder designed for use by one person.
> 10. Keep the ladder and all its steps clean—free of grease, oil, mud, etc.—in order to avoid a fall and possible injury.

FIGURE 16.7 Working safely on a ladder.

 Don't Do This! Don't allow dirt to be piled close to an open ditch or trench. At a minimum, keep piled dirt at least 2 feet from the edge of an open excavation. Ideally, the dirt should be laid off at an angle away from the excavation to avoid accidental backfilling.

It's a good idea to have someone hold a ladder for you, but this is not always practical. We all know that job sites are not a perfect world to work in. I must admit, I have used ladders in mud and other situations when, if I followed tight safety rules, I would not have used them. We, as tradespeople, tend to do what it takes to get the job done. I won't waste my words telling you to follow all of the safety rules that should be followed. After decades in the business, I know that they would be wasted words. But, use common sense and don't push the envelope too far.

Scaffolding Safety

Scaffolding safety is a big issue. Whether you are on pump jacks, swinging scaffolding, rolling scaffolding, or stacked scaffolding, safety is a major issue (Figure 16.8). I have worked on all of these types of scaffolding, and I can't say I have ever enjoyed it. Swinging scaffolding was the worst for me. I prefer to have my feet on the ground. An entire book could be written on the subject of scaffolding safety. There simply isn't room for an exhaustive directive on the subject here, so we will cut to the chase.

Scaffolding on wheels is not that different from standard scaffolding, except for the wheels. Rolling scaffolding has wheel brakes. Use them. Failure to lock the wheels on rolling scaffolding can result in falls. If the scaffolding is hit by equipment, people or some other device, a worker on the scaffolding can lose balance and fall to the floor below. Aside from locking the wheel brakes, there are no special requirements for mobile scaffolding that do not apply to fixed scaffolding.

Pump Jacks

Pump jacks are a common tool for siding contractors. Some contractors use manufactured walk-boards. Other contractors use typical wood boards. I strongly recommend using rated, metal

---

Safety on Rolling Scaffolds

1. Do not lay tools or other materials on the floor of the scaffold. They can easily move and you could trip over them, or they might fall, hitting someone on the ground.
2. Do not move a scaffold while you are on it.
3. Always lock the wheels when the scaffold is positioned and you are using it.
4. Always keep the scaffold level to maintain a steady platform on which to work.
5. Take no shortcuts. Be watchful at all times and be prepared for any emergencies.

FIGURE 16.8 Safety on rolling scaffolds.

walk-boards. They cost more, but they are safer. How much is your health worth? Can you make a living with a broken back or two broken legs? Don't cut corners on equipment that could put you, or your employees, at risk. Any savings you enjoy are likely to be lost down the road. It's a matter of playing the odds, and the odds are that someone will lose. If it is not you, it may be someone who will have a strong basis for a lawsuit against you for inadequate equipment. Don't do it!

Swinging Scaffolds

I hate working from swinging scaffolds, but they are necessary tools in the construction industry. When they are used, they must be rigged properly to ensure worker safety. Don't use swinging scaffolds that are pieced together with makeshift materials. You should not go out on such scaffolding unless it is rigged with the proper gear. Window washers are the crews most often seen on this type of equipment, but painters, plumbers, siding installers and other trades use them. Be careful. Keep wind factors in mind. Personally, I have had enough of this type of work platform, but they can be quite safe for the right individuals. Check all of the rigging before you climb out on the walk board. No job is worth your life, so make sure the rigging is up to the proper standards before you go over the side of a building.

Stationary Scaffolding

Stationary scaffolding can be extremely safe. Does this mean that you should climb towards the sun without inspecting the setup of the scaffolding? I don't think so. It's your life that is at stake.

Don't Do This! Don't use substandard materials when building scaffolding equipment. Any money that you may save for some period of time is likely to be lost several times over when you have a serious accident. If it is not you, it may be one of your employees. It's not worth it. Use scheduled and rated equipment to protect yourself and your workers.

Who set up the scaffolding? Compare it to someone packing a parachute for a paratrooper. Do you trust the installation? Are you going to check the cross braces and the nuts securing them? I doubt it. And, you probably don't have the time or choice. Accidents with fixed scaffolding are more numerous than you might imagine. I could give you the step-by-step checklist for deciding if you are going to climb, but the reality is that you won't be given the opportunity to check every bolt and walk board. Use common sense. If you feel something is not right, question it. Are the walk boards locked into place? Is the boss using boards laid across the scaffolding frame? If so, be especially alert. This is a dangerous practice. I know you have to make a living and have to do things that you are not comfortable with. At least check to see that cross braces are in place and that the work service seems secure. You have to make your own choice of what is worth the risk and what is not. Ideally, you should have a metal walk board that is locked into place on a solid, level platform.

FIRE

Fire claims many lives in many forms. While fire doesn't claim a lot of lives on job sites, it does result in substantial property damages. If you don't have liability insurance, and you should have it, the cost hits hard. Contractors who do have insurance see their insurance premiums shoot up after a fire, assuming that their coverage is not cancelled after a fire mishap.

Remodelers don't normally face many risks of fire. With the exception of burning fires for warmth, which isn't usually required in remodeling work, most remodelers don't run much of a fire risk. Plumbers, who use torches and open flames, are much more likely to experience a fire risk. However, remodeling con-

 Don't Do This! It may be tempting to cut some corners and to use standard, wood boards as walk boards on scaffolding. Don't do it. Invest in metal walk board that will lock into the scaffolding frame and provide good foot traction. Any money you save with wooden boards will be lost with your first fall.

> **To Prevent Fires**
>
> 1. Always keep fire extinguishers handy, and be sure that the extinguisher is full and that you know how to use it quickly.
> 2. Be sure to disconnect and bleed all hoses and regulators used in welding, brazing, soldering, etc.
> 3. Store cylinders of acetylene, propane, oxygen, and similar substances in an upright position in a well-vented area.
> 4. Operate all air acetylene, welding, soldering, and related equipment according to the manufacturer's directions.
> 5. Do not use propane torches or other similar equipment near material that can easily catch fire.
> 6. Be careful at all times. Be prepared for the worst, and be ready to act.

FIGURE 16.9 Tips for fire prevention.

tractors who subcontract plumbers are at some risk, due to the actions of the plumbers. Of all the trades, plumbers are the most likely to create a fire risk. Maintain a strict policy on the use of torches. Have fire extinguishers available on the job site, and make sure that anyone using fire-producing equipment has an insurance binder on file with your office (Figure 16.9).

As a day nears its end, safety is the key factor. Money is important. A good job is essential. Safety must be considered the ultimate goal. You cannot end a good day without it being a safe day, so be alert, be smart, be safe, and be profitable.

QUESTIONS AND ANSWERS

Q: I hate hard hats. Do you feel that they are really needed on residential jobs where the work is a single-family residence?

A: *Hard hats are an added safety precaution. They are often required on large jobs. I keep one in my truck, but I don't always wear it. Since remodeling jobs are not commonly controlled by OSHA regulations, you must use your own common sense.*

Q: What are your feelings on eye protection?

A: *Your eyes cannot be replaced. I feel that safety glasses are essential gear in the toolbox that should be worn at appropriate times.*

Q: How do you feel about transporting material on pipe racks?

A: *There is a level of risk to transporting material on a rack, but if the load is secured properly, there should not be a problem. It is usually when shortcuts are taken that problems occur.*

Q: What do you think the most dangerous part of a job is?

A: *Frankly, I think that driving to the job is the most dangerous part of working. Aside from that, ladder work and trench work are high on my list.*

INDEX

A

Adding new bathrooms, 131-136
  inspections, 142-144
Acoustic paint, 43
Alkyd paint, 42
Appliances, 72
  rip-outs, 335
Asbestos-containing building products, Figure 3.1
Attachment
  deck to existing home, 295-296
  garages, 272-273
  room additions, 150
  screened porch to existing home, 295-296
Attic conversions, 215-216
  garage, 288
  rip-outs, 329-330
  types, 216-217
Awning windows, 200

B

Basements, types of, 2-3
  bathrooms, 15-16
  buried, 3
  daylight, 3
  planning a conversion, 3-4, 11
  rip-outs, 329
  walk-out, 2
Basement conversions, planning, 3-4, 11
Bathrooms, 317
  adding a new, 131-136
  basement, 15-16
  converted porch, 267-268
  inspections, 142-144
  remodeling, 111-112, 113-115, 128-130
  rip-outs, 330-333
Bathing units, 123-124
  bathtub removal, 332
Bedrooms, converted porch, 267
Bidets, 122-123
  rip-outs, 333
Boards (for wall covering), 37
Buried basements, 3

C

Cabinets, 72-109
  custom, 73-74
  existing, 108
  installing, 108-109
  options, 72
  rip-outs, 334-335
  stock, 74
Cabinet options, 74
Carpentry, bathroom, 134
Carpet, 44-45
Carpet padding, 45-47
Casement windows, 200
Ceilings, 316
  basement conversions, 11
  garage, 287
  gazebos, 310-311
  kitchen, 68
  rip-outs, 334
Cinderblock, 314
Clean job sites, 326

Closing in a porch, 262-264
Clothing and jewelry, 339-340
Cosmetic remodeling, 313-314
  other types, 318
  worth, 319-320
Counters, kitchen, 109
  rip-outs, 334-335
Coverings, roof, 177
Crawlspaces, 153
Custom cabinets, 73-74

D

Dark rooms, converted porch, 268
Daylight basements, 3
Debris, dust control and removal, 321, 323
Decking, 298-304
  finishes, 305
Decks, 291-292
  benches and planters, 304-305
  railing, 304-305
Ditches and trenches, safety, 344-345
Doors
  garages, 286-287
  gazebos, 309
  home office, converted porch, 266
  installation, 180
  sliding, 195-1997
  sunrooms, 209-210
Dormer
  benefits, 258
  framing, 249-252
  gable, cutting in, 246-247
  installations, 243-244
  shed, cutting in, 247-249
Double hung windows, 200
Drains, 138
Drywall, 35
Ductwork, attic conversion, 232

Dust and debris control, 321, 324-325

E

Electrical wiring
  adding new bathrooms, 134
  attic conversions, 233
  basement bathrooms, 28-29
  gazebos, 308
  kitchens, 68-69
  sunrooms, 211
Enclosed porch conversions, 261-262
  defining the use, 269
  type of use, 269
Exterior finishes, garage, 285

F

Faucets
  bathroom, 127
  kitchen, 71
Finger-joint trim, 56
Finish work, attic conversion, 241
Fire, safety, 350-351
Fixed glass, sunrooms, 197
Fixtures
  light, rip-out, 333
  placement in basement bathrooms, 22
  plumbing, bathroom, 115-128
  replacing existing in bathrooms, 112-113
Flashing, 178
Floor
  attic conversion, structure, 222-229
  bathroom rip-outs, 331
  finish for garage, 277
  framing, 156-161
  protection of existing, 324

sheathing, 161-161
Flooring, 33-35, 44, 316-317
  basement conversions, 7
  carpet, 44-45
  padding, 45-47
  garage, 288
  gazebos, 311
  home office, converted porch, 266
  kitchen, 67
  rip-outs, 334
  rough, attic conversion, 234
  sunrooms, 212
  tile, 48-49
  vinyl, 47-48
  wood, 49-56
Footings
  garages, 275-277
  room additions, 152
Foundation
  cinderblock, 314
  cosmetic remodeling, 314-315
  crawlspaces, 153
  decks, 293-295
  footings, 152
  full, 153-155
  gazebos, 293-295
  pier, 156, 314-315
  room additions, 150, 152-155
  screened porches, 293-295
  sunroom, adding, 188
Framing
  dormer, 249-252
  floor, 156-161
  garage door, 286-287
  general, garage, 277-282
  sunroom, 188
    walls, 189-193
    windows, 201-208
  temporary bracing, porch, 263-264
  wall, 162-166
Full foundations, 153-155

G

Gable dormer, cutting in, 246-247
Garages, 271-272
  attics, 288
  converting to living space, 285-286
  detached or attached, 272-273
Gazebos, 291-292, 307
  walls, 307-308

H

Hardboard
  paneling, 37
  siding, 183
Headroom, attic conversion, 234-236
Heating
  gazebos, 309-310
  kitchens, 71
  sunrooms, 212
Home office, converted porch, 265-276

I

Ice dams, 178, 180
Inlaid sheet vinyl flooring, 47
Inspections, 29
  adding a new bathroom, 142-144
Installing cabinets, 108-109
Insulation
  attic conversion, 239-241
  garage, 284-285, 287
  porch, 264
Interior framing, 9-10, Figure1.3
Interior trim, for wood flooring, 54-56
Interior walls, 315-316

J

Jewelry, clothing and, 339-340
Joist structure
  decks, 296-298
  screened porches, 296-298

K

Kitchen
  ceilings, 68
  counters, 109
  expansion, 61-65
  floors, 67
  remodeling, 59-60, 1009, 317-318
  rip-outs, 334-335
Kitchen counters, 109
  rip-outs, 334-335
Kitchen expansion, 61-65

L

Ladders and scaffolds, safety, 346-350
Latex paints, 41
Laundry rooms, converted porch, 268
  noise reduction, 268-269
Lavatories, 115-117, Figure 5.4, Figure 5.5
  rip-out, 332-333
Lighting
  home office, converted porch, 266
  rip-out, bathroom fixtures, 333
Living space
  rip-out problems, 328

M

Manipulating walls, 65-66

Mechanical installations
  attic conversions, 237, 239
  garages, 282, 284, 287
Messy job-sites, 322-323

N

Noise reduction, laundry room, 268-269

P

Paint, 40-43
  acoustic, 43
  alkyd, 42
  dormers, 257
  latex, 41
  polyurethane, 42
  textured, 42-43
  urethane, 42
Paint or paper, wallcoverings, 37-38
Paneling, 35, 37
Partitions, basement bathroom, 28
Pier foundations, 156, 314-315
Piping, attic conversions, 233
Pitch, roof dormer, 249, 251-252
Plank flooring, 54
Plumbing
  adding a new bathroom, 134-142
  bathroom fixtures, 115-128
  gazebos, 310
  home office, converted porch, 266
  kitchens, 69-71
  sunrooms, 212
  vents, attic conversion, 229
Plywood paneling, 37
Polyurethane paint, 42

R

Railing, 304-305
Replacing existing fixtures, 112-113
Rip-outs of existing conditions, 327-328
Ripping out, 60-61
Roof
  coverings, 177
  gazebos, 309
  materials, 180
  pitch, dormer, 249, 251-252
  sheathing, 176-177
  stick-built, 173-176
  structures, 170-176
  sunroom, 193-194
  trusses, 171-176
Roofing materials, 177, 180
Room additions, 147-148
Rotovinyls, 48
Roughed in plumbing, 17-18

S

Safety issues, 337-339
Scaffolds, safety, 348-350
Screened porches, 291-292, 311
Sewage ejectors, 23
Shake siding, 183
Sheathing
  floor, 161-162
  roof, 176-177
  wall 166-169
Shed dormer, cutting in, 247-249
Shelves, built-in, 267
Shower bases, rip-outs, 332
Siding
  dormer, 252-257
  gazebos, 309
  hardboard, 183
  room addition, 180-184
  shake, 183,
  vinyl, 182
  wood, 184-185
Simple conversions, porch, 262
Sinks, kitchen, 70-71
Site visit, room additions, 148-149
Skylights, 71-72
  bubble, 194
Sliding doors, 195-197
Sliding windows, 200
Soil conditions, 272-275
Soil loads, 152
Solarium panels, 194-195
Stairs, 6
attic conversion, 217-222
Steps
  deck, 305
  porch, 264, 305
Stick-built roofs, 172-176
Stock cabinets, 74
Stretch cushioned vinyl, 48
Sunrooms, adding, 187-188
  accessories, 213
Support columns, 8-9

T

Temporary bracing, porch framing, 263-264
Textured paint, 42-43
Tile
  bathroom walls, rip-out, 331
  flooring, 43-44, 48-49
  tub surrounds, rip-out, 331-332
Toilets, 117, 121-122
  rip-outs, 33
Tools, safety, 340-344
Trenches, ditches and, safety, 344-345
Trim, dormer, 257
Trusses, roof, 171-172
Tub surrounds, tile, 331-332

U

Underground obstacles, 293
Urethane paint, 42
Using existing cabinets, 108

V

Vanities, rip-out, 332-333
Ventilation, roof, 178
Vents
  adding new bathroom, 139
  plumbing, attic conversion
Venting, 20-21
Vinyl flooring, 47-48
  inlaid sheet, 47
  rotovinyls, 48
  stretch cushioned vinyl, 48
Vinyl siding, 182-183

W

Walk-out basements, 2
Walls
  attic conversion, 237
  basement conversions, 6-7
  framing, 162-166
  gazebos, 307-309
  interior, 315-316
  kitchen, 67-68
    rip-outs, 334
  manipulating in kitchens, 65-66
  preparation, 6-7
sheathing, 166-169
surfaces, home office, 267
Wall coverings, 33-35
  basement conversions, 11
  garage, 285, 287
  paint, 40-43
  paint or paper, 37-38
  paneling, 35, 37
  tile, 43-44
  wallpaper, 38-40
Water piping, 21-22
  adding new bathroom, 139-142
Waterproofing a basement, 4-6
Weather protection, dormers, 244, 246
Whirlpool tubs, 125-127
Windows
  attic conversions, 237
  awning, 200
  casement, 200
dormer, 252
  double hung, 200
  framing, sunroom, 201-208
  gazebos, 309
  installation, sunroom, 180, 201-208
  kitchen, 71-72
  place, home office, 267
  roof windows, 194
  sliding, 200
  sunroom, 197-201
Wiring (see Electric wiring)
Wood flooring, 49-56
  interior trim, 54-56
  plank, 54
  strip, 50-53
  tiles, 54
Wood siding, room additions, 184-185
Wood tiles, 54

Z

Zoning
  adding a sunroom, 188
  decks, 292-293
  garages, 273
  gazebos, 292-293
  room additions, 149-150
  screened porches, 292-293